# HERBICIDE RESISTANCE IN WEEDS AND CROPS

Edited by **Zvonko Pacanoski**

**Herbicide Resistance in Weeds and Crops**
http://dx.doi.org/10.5772/65569
Edited by Zvonko Pacanoski

**Contributors**

Kassio Mendes, Miriam Inoue, Valdemar Tornisielo, Ricardo Silveira, André Andres, Germani Concenço, Fábio Schreiber, Dirceu Agostinetto, Leandro Vargas, João Pedro Oliveira Behenck, Giovanni A. Caputo, Ygor Sulbach, Mohammad Alebrahim, W. James Grichar, Sava Vrbničanin, Dragana Bozic, Danijela Pavlovic, Zvonko Pacanoski

**Published by InTech**
Janeza Trdine 9, 51000 Rijeka, Croatia



**Publishing Process Manager** Nina Kalinic
**Technical Editor** SPi Global
**Cover** InTech Design team

First published October, 2017
Printed in Croatia
Legal deposit, Croatia: National and University Library in Zagreb

Additional hard copies can be obtained from orders@intechopen.com

Herbicide Resistance in Weeds and Crops, Edited by Zvonko Pacanoski
p. cm.
Print ISBN 978-953-51-3555-5
Online ISBN 978-953-51-3556-2

For EU product safety concerns:
IN TECH d.o.o., Prolaz Marije Krucifikse Kozulić 3, 51000 Rijeka, Croatia,
info@intechopen.com or visit our website at intechopen.com.

# Meet the editor

Prof. Dr. Sci. Zvonko Pacanoski is an associated professor at the Faculty of Agricultural Sciences and Food, Institute for Plant Protection in Skopje, Republic of Macedonia. He graduated in 1998 at the Faculty for Agriculture in Skopje. He had his master's thesis in 2003 and dissertation in 2008 at the same faculty. He participated at the different training programs and courses in the Netherlands, Germany, Switzerland, and Greece, and in 2009, he specialized in Sustainable Agriculture and Rural Development (SARD) in CIHAEM, Bari, Italy.

# Contents

# Preface

Arable weeds have been the major biotic cause of crop yield losses since the origins of agriculture. Weeds result in 34% loss of crop yield, on average, worldwide. Herbicides are by far the most effective weed control tools ever developed, controlling 90 to >99% of the weeds targeted. Consequently, the arable surface treated and range of weed species targeted by herbicides increased rapidly worldwide after their development. Weed communities have evolved over time in response to this control practice imposed on them. For the past half century, a principal method of weed management in commercial crops in the most developed countries has been the use of herbicides. But, this golden age of herbicides was quickly cut short, however, by the detection of the first herbicide-resistant weeds in the early 1970s, although it was described as a potential problem as early as 1957. Resistant weeds have been evolving worldwide from selection pressure caused by the repeated use of herbicides with the same mechanism of action in conventional crop cultivars. Today, herbicide resistance has been reported in 251 weed species (146 dicots and 105 monocots) in more than 670,000 fields in 90 crops in 66 countries. Resistance has been reported to all major known herbicide modes of action, and no new mode of action has been marketed since 1991.

Herbicide resistance in weeds is a global problem. Resistance to herbicides in arable weeds is increasing rapidly worldwide and threatening global food security. Resistance has now been reported to all major herbicide modes of action despite the development of resistance management strategies in the 1990s.

From the other side, development of herbicide-resistant crops has resulted in significant changes to agronomic practices, one of which is the adoption of effective, simple, low-risk, crop production systems with less dependency on tillage and lower energy requirements. Overall, the changes have had a positive environmental effect by reducing soil erosion, the fuel use for tillage, and the number of herbicides with groundwater advisories as well as a slight reduction in the overall environmental impact quotient of herbicide use. However, herbicides exert a high selection pressure on weed populations and density and diversity of weed community's change over time in response to herbicides and other control practices imposed on them.

This book focuses on the recent progress made in understanding the genetic and evolutionary mechanisms underlying herbicide resistance in weeds. Current controversies on key aspects of resistance evolution are discussed. The authors of *Herbicide Resistance in Weeds and Crops* highlight crucial present and future research directions and challenges connected with understanding of weed resistance development and the importance and impact of herbicide-resistant crops.

The information provided in this book serves as a beneficial device to illustrate current herbicide resistance research touching agriculture and environment, as well. *Herbicide Resistance in Weeds and Crops* should be principally valuable for scientists and researchers interested in advancing research strategies concentrated on accepting weed resistance as a global problem and proactive, evolutionary-based weed management options for agriculture today. I hope that this book will provide the scientific community with a source of crucial research knowledge to assist format prospective research and understanding weed and crop herbicide resistance.

**Dr. Zvonko Pacanoski**
Associate Professor of Weed Science,
Institute for Plant Protection,
Faculty of Agricultural Sciences and Food,
Skopje, Republic of Macedonia

Chapter 1

# Introductory Chapter: Actual Issues (Moments) in Herbicide Resistance Weeds and Crops

Zvonko Pacanoski

Additional information is available at the end of the chapter

http://dx.doi.org/10.5772/67646

Weeds are the most important pest complexes that globally reduce the ability of mankind to produce food, fiber, and fuel [1]. They have always been a component of agriculture and have successfully thwarted all attempts by humans to eliminate them from interfering with crop production [2]. Weeds result in a 34% loss of crop yield, on an average, worldwide.

In most parts of the world, herbicides are the dominant technology and the most effective weed control tools ever developed for the control of weeds that infest crops, killing 90 to >99% of the weeds targeted [3]. Consequently, the arable surface treated and the range of weed species targeted by herbicides increased rapidly worldwide after their development [1]. Weed communities have evolved over time in response to this control practice imposed on them. For the past half century, a principal method of weed management in commercial crops has been using herbicides in the most developed countries. But, this golden age of herbicides was quickly cut short, by the detection of the first herbicide-resistant weeds in the early 1970s [4], although it was described as a potential problem as early as 1957 [5, 6].

Resistance of weeds to herbicides is an unwanted response formed after the remade use of a same herbicide, where a weed population is once controlled with the similar efficacy by an herbicide which, in regular circumstances in an appropriate crop, had been efficient against the weed population [7]. Over the last several decades, in situations of intense herbicide usage, there have been many examples of the evolution of weed populations that are resistant to herbicides [8, 9]. However, "resistance evolution" does not mean that an herbicide directly changes a plant genetically (i.e., by causing mutations). Instead, the herbicide selects plants with some level of natural genetic resistance to the mechanism of action.

Weed adaptations to management tactics, including the biochemical mimicry in the form of an evolved resistance to the herbicides that are used for weed control, have increased rapidly throughout agriculture and now threaten global food security [10, 11]. From an evolutionary perspective, many factors influence the dynamics of herbicide-resistant evolution under herbicide selection [12, 13]. One crucial factor in herbicide-resistant evolution is the selection pressure caused by the repeated use of herbicides with the same mechanism of action in conventional crop cultivars [14], of which a major determinant is the herbicide use rate [15]. The use of an herbicide (or herbicides from the same herbicide group) continuously for many years can drastically decrease the number of susceptible biotypes within the natural weed population and dramatically increase the number of resistant biotypes.

Resistance has increased rapidly since 1975, and today, there are currently 477 unique cases (species × site of action) of herbicide-resistant weeds globally, with 251 species (146 dicots and 105 monocots) in more than one million fields. Herbicide-resistant weeds have been reported in 90 crops in 66 countries [8]. The total area affected, although not estimated, may cover several thousand hectares of crops regularly treated with herbicides in countries such as Australia, Canada and the United States of America, as well as countries in the European Union and South America. Weeds have developed resistance to 23 of the 26 known herbicide sites of action and to 161 different herbicides [8], and no new mode of action has been marketed since 1991 [16].

Herbicide resistance in weeds is a global problem. Resistance to herbicides in arable weeds is increasing rapidly worldwide and threatening global food security. Resistance has now been reported to all major herbicide modes of action despite the development of resistance management strategies in the 1990s. Despite it being a known issue, farmers in many states reveal the problem of herbicide weakness when the resistance is present in the field; alike bad, occasionally, they are using other herbicide ingredients that have the same mechanism of action as the one already used, which deteriorates the problem.

Proactive, evolutionary-based weed management options that integrate both herbicides and non-chemical tools are of utmost importance in agriculture today [14]. As resistance is generally the consequence of using a single herbicide repeatedly, any proactive or reactive approach should take an opposite view: the use of a diverse method to avoid repetition as much as possible. Because of that, herbicide-resistant weed management practices most often recommended by weed scientists include (1) identification of resistant populations through diligent field monitoring; (2) biosanitary practices, such as cleaning equipment and removing and destroying resistant plants to prevent re-infestation of the field with resistant seeds or plant parts; (3) crop rotations and/or the use of competitive covers that allow the use of alternative mechanism of actions or that change the balance of weeds in a field or both; (4) cultivation and hoeing that provide weed control, which reduces reliance on herbicides; (5) using herbicide rotations and mixtures, which include compounds from classes of herbicides with different modes of action that control similar spectra of weeds; (6) using only labeled herbicide rates at labeled application timings; (7) introduction of new herbicides and herbicide modes of action to replace those herbicides failing due to resistance; and (8) controlling weed escapes [17].

Since 1996, herbicide-resistant crops (HRCs) have had a major effect on agriculture, particularly in the United States of America, Brazil, Argentina, and Canada [18]. The introduction of HRCs in the United States of America, for example, helped solve a major weed-management problem that was developing at that time—the evolution of weeds resistant to the acetolactate synthase (ALS)-inhibiting and protoporphyrinogen oxidase (PPO)-inhibiting herbicides [19]. The adoption of HRCs has resulted in significant changes to agronomic practices as well. HRCs have allowed for the acceptance of practical, uncomplicated, and below hazard crop production systems with minor dependencies on soil cultivation and diminished energy demands [20]. Long-term differences have had an affirmative environmental issue by diminishing soil erosion [21], fueling the needs for soil cultivation [22], and numbering herbicides with groundwater advisories [23], leading to a slight reduction in the overall environmental impact quotient of herbicide use [24, 25]. Because of the adoption of herbicide-resistant crops, conservation tillage used in crop production has increased [18, 26, 27], and the volume of herbicides used in HRCs has decreased [28]. Finally, the effect on soil and plant microbial populations has not been shown to be a potential environmental risk [29, 30]. Because of these reasons by 2015, more than 179 million hectares worldwide were planted to HR varieties of soybean, maize, canola, cotton, alfalfa, and sugar beets [31].

Controversies surrounding HRCs commonly focus on human and environmental safety, labeling and consumer choice, intellectual property rights, ethics, food security, poverty reduction, and environmental conservation [32]. Of potential concern with HRCs, it is a possibility for the development of weed-resistant mechanisms to non-selective herbicides [21] and shifts in the composition of weed flora, which provoke a change of biodiversity [33, 34]. Other risks for this system are "volunteer HR crops". Volunteer plants of the previous HR crop in the next HR crop can be a problem if the next HR crop is resistant to the same herbicide like the previous HR crop [35]. Possible direct influences of HRCs acceptance on biodiversity, especially in South America, encompassed a shift in the genetic diversity of crops, expended volunteer crop issues, and induced aggression by resistant varieties of natural ranges above the farm line [36]. Also, there is a risk correlated with the probability for the exchange of genetic material between related HRCs, from the one side wild progenitors and conventional crops and weeds from the other side [37, 14].

## Author details

Zvonko Pacanoski

Address all correspondence to: zvonko_lav@yahoo.com

Faculty of Agricultural Sciences and Food, Institute for Plant Protection, Skopje, The Republic of Macedonia

## References

[1] Oerke EC. Crop losses to pests. Journal of Agricultural Science 2006; **144**: 31-43.

[2] Vigueira CC, Olsen KM, Caicedo AL. The red queen in corn: agricultural weeds as models of rapid adaptive evolution. Heredity 2013; **110**: 303-311.

[3] Foster DK, Ward P, Hewson RT. Selective grass-weed control in wheat and barley based on the safener fenchlorazole-ethyl. In British Crop Protection Council (ed.). Proceedings of the Brighton Crop Protection Conference – Weeds. British Crop Protection Council; 1993, pp. 1267-1272.

[4] Ryan GF. Resistance of common groundsel to simazine and atrazine. Weed Science 1970; **18**: 614-616.

[5] Hilton HW. Herbicide tolerant strains of weeds. In Hawaiian Sugar Planters' Association Annual Report. Honolulu, HI: University of Hawaii, Manoa Library; 1957, 69 p.

[6] Switzer CM. The existence of 2,4-D resistant wild carrot. Proceedings North Central Weed Science Society 1957; **9**: 75-77.

[7] Taberner A, Ranzenberger PAC, Larios CZ. Management of herbicide-resistant weed populations-100 questions on resistance. Rome: Food and Agriculture Organization of the United Nations; 2008.

[8] Heap I. The international survey of herbicide resistant weeds. December 9, 2016. Available from: www.weedscience.org

[9] Powles SB, Yu Q. Evolution in action: plants resistant to herbicides. Annual Review of Plant Biology 2010; **61**: 317-347.

[10] Delye C, Jasieniuk M, Le Corre V. Deciphering the evolution of herbicide resistance in weeds. Trends Genetics 2013; **29**: 649-658.

[11] McElroy JS. Vavilovian mimicry: Nikolai Vavilov and his little-known impact on weed science. Weed Science 2014; **62**: 207-216.

[12] Darmency H. Genetics of herbicide resistance in weeds. In SB Powles and JAM Holtum (eds). Herbicide resistance in plants: biology and biochemistry. Boca Raton, FL: Lewis; 1994, pp. 263-295.

[13] Jasieniuk M, BruleBabel AL, Morrison IN. The evolution and genetics of herbicide resistance in weeds. Weed Science 1996; **44**: 176-193.

[14] Vencill WK, Nichols RL, Webster TM, Soteres JK, Mallory-Smith C, Burgos NR, Johnson WG, McClelland MR. Herbicide resistance: toward an understanding of resistance development and the impact of herbicide-resistant crops. Weed Science 2012; 60(sp1): 2-30.

[15] Manalil S, Busi R, Renton M, Powles SB. Rapid evolution of herbicide resistance by low herbicide dosages. Weed Science 2011; **59**: 210-217.

[16] Duke SO. Why have no new herbicide modes of action appeared in recent years? Pest Management Science 2012; **68**: 505-512.

[17] Owen MDK. Diverse approaches to herbicide-resistant weed management. Weed Science 2016; **64**(sp1): 570-584.

[18] Brookes G, Barfoot P. Global impact of biotech crops: environmental effects 1996-2009. GM Crops Food 2011; **2**: 1-16.

[19] Norsworthy JK, Ward SM, Shaw DR, Llewellyn RS, Nichols RL, Webster TM, Bradley KW, Frisvold G, Powles SB, Burgos NR, Witt WW, Barrett M. Reducing the risks of herbicide resistance: best management practices and recommendations. Weed Science 2012; **60**(sp1): 31-62.

[20] Carpenter J, Gianessi L. Herbicide tolerant soybeans: why growers are adopting roundup ready varieties. Agbioforum 1999; **2**: 65-72. Available from: http://www.agbioforum.org/v2n2/v2n2a02-carpenter.htm

[21] Fawcett R, Towery D. Conservation tillage and plant biotechnology: how new technologies can improve the environment by reducing the need to plow. West Lafayette, IN: Conservation Technology Information Center; 2002. Available from: http://croplife.intraspin.com/Biotech/papers/35%20Fawcett.pdf

[22] Duke SO, Powles SB. Glyphosate-resistant crops and weeds: now and in the future. Agbioforum 2009; **12**: 346-357.

[23] Krueger RW. The public debate on agrobiotechnology: a biotech company's perspective. Agbioforum 2001; **4**: 209-220.

[24] Foresman C, Glasgow L. US grower perceptions and experiences with glyphosate-resistant weeds. Pest Management Science 2008; **64**: 388-391.

[25] Young BG. Changes in herbicide use patterns and production practices resulting from glyphosate-resistant crops. Weed Technology 2006; **20**: 301-307.

[26] Cerdeira AL, Duke SO. The current status and environmental impacts of glyphosate-resistant crops: a review. Journal of Environmental Quality 2006; **35**: 1633-1658.

[27] Dill GM, CaJacob CA, Padgette SR. Glyphosate-resistant crops: adoption, uses and future considerations. Pest Management Science 2008; **64**: 326-331.

[28] Brookes G, Barfoot P. GM crops: global socio-economic and environmental impacts 1996-2007. Dorchester, UK: PG Economics; 2009.

[29] Dunfield KE, Germida JJ. Impact of genetically modified crops on soil- and plant-associated microbial communities. Journal of Environmental Quality 2004; **33**: 806-815.

[30] [FAO] Food and Agriculture Organization of the United Nations. Report of the FAO expert consultation on environmental effects of genetically modified crops. 2003. Available from: http://www.fao.org/DOCREP/FIELD/006/AD690E/AD690E05.htm

[31] James C. Global status of commercialized biotech/GM crops: 2015 (ISAAA Brief 51-2015). Ithaca, NY: International Service for the Acquisition of Agri-biotech Applications (ISAAA); 2015.

[32] Pacanoski Z. Herbicide-resistant crops, advantages and risks. Herbologia, Saraevo 2006; VII(1): 47-57.

[33] Spahillari M, Hammer K, Gladis T, Diederichsen A. Weeds as part of agrobiodiversity. Outlook Agriculture 1999; **28**: 227-232.

[34] Duke SO, Cerdeira AL. Potential environmental impacts of herbicide-resistant crops. Collection of Biosafety Reviews 2005; **2**: 66-143.

[35] Beckie HJ, Owen MDK. Review: herbicide-resistant crops as weeds in North America. CAB Reviews: Perspectives in Agriculture, Veterinary Science, Nutrition and Natural Resources 2007; **2**(044).

[36] Riches CR, Valverde BE. Agricultural and biological diversity in Latin America: implications for development, testing, and commercialization of herbicide-resistant crops. Weed Technology 2002; **16**(1): 200-214.

[37] Mallory-Smith CA, Olguin ES. Gene flow from herbicide-resistant crops: it's not just for transgenes. Journal of Agriculture Food and Chemistry 2011; **59**(11): 5813-5818

Chapter 2

# Weed Resistance to Herbicides

Sava Vrbničanin, Danijela Pavlović and Dragana Božić

Additional information is available at the end of the chapter

http://dx.doi.org/10.5772/67979

**Abstract**

Unfortunately, herbicide resistance developed shortly after the introduction of the herbicides 2,4-D in 1957. According the herbicide resistance mechanisms, all processes can be grouped as follows: target-site resistance, non-target-site resistance, cross-resistance and multiple-resistance. Target-site resistance is generally due to a single or several mutations in the gene encoding the herbicide-target enzyme, which, in turn, decreases the affinity for herbicide binding to that enzyme. Non-target-site resistance is caused by mechanisms that reduce the amount of herbicidal active compound before it can attack the plant through the reduced absorption or altered translocation, increased herbicide sequestration or enhanced herbicide metabolism. Cross-resistance means that a single-resistance mechanism causes resistance to several herbicides with some mode of action. Multiple-resistance is a situation where two or more resistance mechanisms are present within the same plant, often due to sequential selection by herbicides with different modes of action. Currently, herbicide resistance has been reported in 478 weed biotypes (252 weed species) in 67 countries. Many of those biotypes are resistant to acetolactate synthase (ALS) inhibitors, PS II inhibitors, ACC-ase inhibitors and EPSPS inhibitors. Strategy for herbicide-resistance weed management must involve all the available preventive, cultural, mechanical and chemical measures for effective, safe and cost-effective weed control.

**Keywords:** weed, herbicide, resistance, management

## 1. Introduction and general overview of resistance

Since the introduction of 2,4-D as a first selective herbicide in 1947, herbicides have had a major positive impact on weed management in all over the world. Unfortunately, herbicide resistance developed shortly after the introduction of the herbicides. The phenomenon of resistance can be defined as the decreased response of a species' population to herbicide [1].

It is also defined as a 'survival of a segment of the population of a weed species following an herbicide dose lethal to the normal population' [2]. In addition, resistance can be defined as 'the inherited ability to survive treatment by a herbicide' [3], or it is a 'phenomen which occurs as a result of heritable changes to biochemical processes that enable weed species survival when treated with a herbicide' [4].

Weed resistance to herbicides is a normal and predictable outcome of natural selection. In that context, rare mutations that confer herbicide resistance exist in wild/weed populations before any herbicide introduction. These mutations increase over time after each herbicide application until they become predominant at what time the weed population is called resistant [5]. The first confirmed herbicide-resistant weed species was *Senecio vulgaris* that had developed resistance to PS II inhibitors (atrazine and simazine) after the herbicides had been applied once or twice annually for 10 years [6]. Therefore, about 30-resistant weed populations have been confirmed within the first decade, mostly in N. America and W. Europe [7]. Some weed species, such as *Lolium rigidum, Echinochloa crus-galli* var. *crus-galli, Poa annua, Alopecurus myosuroides, Echinochloa colona, Eleusine indica, Amaranthus* sp., etc. have a high affinity to develop resistance especially due to their congenital genetic variability. Additionally, herbicides of different chemical groups and different modes of action (e.g. sulfonylurea and synthetic auxins) can greatly differ in their risk levels for resistance. On the other hand, different chemical groups with the same mode of action such as herbicide inhibitors of acetolactate/acetohydroxyacid synthase (ALS/AHAS) (sulfonylurea, pyrimidinyl(thio)benzoate, sulfonylaminocarbonyl-triazolinone, imidazolinone) can also be distinguished in their risk level for resistance.

Currently, herbicide resistance has been reported in 478 weed biotypes (252 weed species) in 67 countries. Many of those biotypes are resistant to ALS inhibitors, B/2 (97 dicots + 62 monocots), PS II inhibitors (C1/5 = 51 + 23, C2/7 = 10 + 18, C3/6 = 3 + 1), ACC-ase inhibitors, A/1 (48 monocots) and EPSPS inhibitors, G (19 + 17). The highest number of confirmed resistant weed species belongs to the families: *Poaceae* (80 species), *Asteraceae* (39), *Brassicaceae* (22), *Cyperaceae* (12), *Amaranthaceae* (11), *Scrophulariaceae* (9), *Chenopodiaceae* (8), *Alismataceae* (7), *Polygonaceae* (7) and *Caryophyllaceae* (6). According to the number of active ingredients (a.i.), those four sites of action participate in the next relation: 50 a.i. from ALS inhibitors, 24 a.i. from PS II inhibitors, 15 a.i. from ACC-ase inhibitors and 2 a.i. from EPSPS inhibitors. Atrazine (PS II inhibitors) is an active ingredient, which was confirmed by the greatest number of weed resistant species (66), the second is imazethapyr (44), followed by tribenuron-methyl (43), imazamox (37), chlorsulfuron (36) metsulfuron-methyl (35), glyphosate (34), iodosulfuron-methyl-sodium (33), fenoxaprop-P-ethyl (31), simazine (31), bensulfuron-methyl (29), thifensulfuron-methyl (27), fluazifop-P-bityl (25), pyrazosulfuron-ethyl (25), etc. In relation to herbicide-resistant weeds by county and site of action top 10 counties are the United States, Australia, Canada, France, Brazil, China, Spain, Israel, Japan and Germany [7] (**Table 1**).

In Serbia, study of weed resistance to herbicides started in the 1990s with resistance of *Amaranthus retroflexus* and *Chenopodium hybridum* to PS II inhibitors (atrazine) [8–10]. Until today, in Serbia, as a small county with less than 3 million ha arable lands, in the last 15 years, eight herbicide-resistant weed species were confirmed: *A. retroflexus, Setaria viridis, C. hybridum* and *Abutilon theophrasti* to PS II inhibitors, as well as *A. retroflexus, E. crus-galli, Datura stramonium, Chenopodium*

http://dx.doi.org/10.5772/67979

| Country | Total number of resistant weed species | Number of resistant weed species according to the site of action | | | |
|---|---|---|---|---|---|
| | | ALS | ACC-ase | PS II | EPSPS |
| USA | 156 | 51 | 15 | 26 | 16 |
| Australia | 84 | 25 | 12 | 7 | 13 |
| Canada | 64 | 25 | 4 | 12 | 5 |
| France | 48 | 16 | 6 | 22 | 2 |
| Brazil | 42 | 19 | 6 | 4 | 8 |
| China | 41 | 14 | 8 | 1 | 2 |
| Spain | 37 | 8 | 2 | 18 | 5 |
| Israel | 36 | 12 | 6 | 12 | 2 |
| Japan | 36 | 21 | 2 | 1 | 3 |
| Germany | 32 | 10 | 5 | 13 | 0 |

**Table 1.** Top 10 countries with the most number of confirmed resistant weed species.

*album* and *Sorghum halepense* to ALS inhibitors [11–19]. According to the herbicide resistance mechanisms, all processes can be grouped as follows: target-site resistance, non-target-site resistance, cross-resistance and multiple-resistance [20–22].

*Target-site resistance* (TSR) is generally due to a single or several mutations in the gene encoding the herbicide-target enzyme, which, in turn, decreases the affinity for herbicide binding to that enzyme. Most, but not all cases of resistance to herbicide ALS inhibitors, ACC-ase, triazine, dinitroaniline etc. are due to modifications of the site of action of the herbicide. In addition, gene overproduction (amplification) is the most recently identified herbicide resistance mechanism, for example, EPSPS gene amplification correlates with glyphosate resistance in *Amaranthus palmeri* and *Kochia scoparia* [23–25], and causes resistance by increasing the production of the target enzyme, effectively diluting the herbicide in relation to the target site (**Figure 1**).

*Non-target-site resistance* **(NTSR)** is caused by mechanisms that reduce the amount of herbicidal active compound before it can attack the plant. Reduced absorption (penetration) or altered translocation, increased herbicide sequestration or enhanced herbicide metabolism (detoxification) can cause resistance due to the restriction of herbicide movement where the herbicide does not reach its site of action in sufficient concentration to cause plant mortality. Active vacuolar or cell walls sequestration can keep the herbicide from the site of action leading to resistance. For example, vacuolar herbicide sequestration correlates with glyphosate resistance in *Conyza canadensis*, *Lolium* sp. etc. [26, 27] (**Figure 1**). Finally, the biochemical reactions that detoxify herbicides can be grouped into four major categories: oxidation, reduction, hydrolysis and conjugation [28].

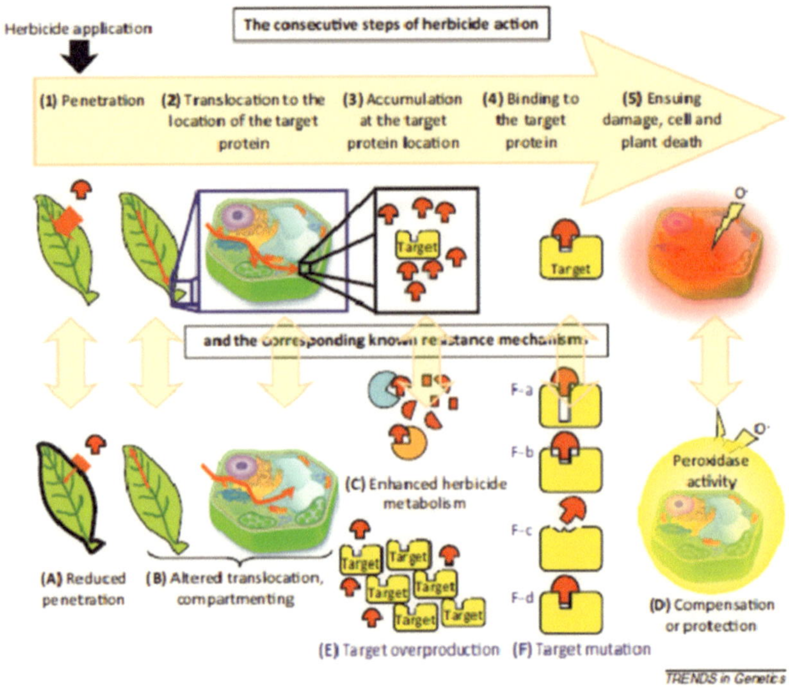


**Figure 1.** The route of the herbicide after the application, and the possible mechanisms of resistance in plant. After application: (1) herbicide absorption/penetration, (2) translocation, (3) accumulate at the target protein location, and (4) binding to the target protein, (5) disruption of the biosynthesis pathways or cell structures, and/or generation of cytotoxic molecules. NTSR mechanisms: (A) reduction in herbicide penetration, (B) altered translocation of the herbicide away from the target protein, (C) enhanced detoxification of the herbicide, or (D) enhanced neutralization of cytotoxic molecules generated by herbicide action. TSR mechanisms: (E) target protein overproduction, and/or (F) structural mutations that modify the 3D structure and electrochemical properties of the target protein. Structural mutations can have no, moderate or strong negative effects on the stability of herbicide binding to the target protein, which results in (F-a) no, (F-b) moderate or (F-c) marked reduction in herbicide sensitivity at the protein level, respectively; or can (F-d) increase the stability of herbicide binding to the target protein, which results in an increase in herbicide sensitivity at the protein level (downloaded from Ref. [32]).

*Cross-resistance* **(CR)** means that a single-resistance mechanism causes resistance to several herbicides. CR can be conferred by a single gene or by two or more genes influencing a single mechanism. There are two types of CR: target-site cross-resistance (TS-CR) and non-target-site cross-resistance (NTS-CR). The most common type of CR is TS-CR where an altered target site confers resistance to many or all of the herbicides that inhibit the same enzyme, for example, Trp-574-Leu amino acid substitution within the ALS gene was found in two populations of *Cyperus iria* after exposition to bispyribac-sodium, halosulfuron, imazamox and penoxsulam [29]. On the other hand, NTS-CR is type of herbicide resistance in which a mechanism other than resistant enzyme target sites is involved (e.g. reduced absorption, translocation, or enhanced herbicide detoxification) [30].

*Multiple-resistance* is a situation where two or more resistance mechanisms are present within the same plant, often due to sequential selection by herbicides with different modes of action (e.g. resistance of *Lolium* sp. populations to glyphosate and ACC-ase inhibitors, as well as resistance to glyphosate and ALS inhibitors were confirmed by multiple-resistance [31]).

http://dx.doi.org/10.5772/67979

## 2. Weed resistance to herbicides photosystem II inhibitors, triazines

The triazine herbicides were discovered in the J.R. Geigy Ltd. laboratories, an international chemical company founded in 1952 and based in Basel, Switzerland [33]. Generally, in the latter half of the twentieth century, triazines have played a significant role in the promotion of the crop production. Atrazine is one of the most used triazine herbicides in agriculture for control of annual monocots (*Setaria* sp., *E. crus-galli, Digitaria sanguinalis*) and dicot weed species (*Amaranthus* sp., *Chenopodium* sp., *Cirsium arvense, D. stramonium, Sonchus* sp., *Xanthium strumarium*, etc.) and is the most widely used herbicide in maize, orchards and sorghum crops. Triazines specifically inhibit photosystem II (PS II) in plants and in all organisms with oxygen-evolving photosystems. Generally, they prevent electron transfer by displacing plastoquinone ($Q_B$) from a specific binding site on the D1 protein subunit of PS II [34, 35].

The intensive use of triazines resulted in two important cases: appearance of atrazine-resistant weed species, leading to the increased use of herbicide mixtures or alternative herbicides. The first confirmed atrazine-resistant weed species [6] helped identify the herbicide-binding D1 protein in PS II. After the psbA gene was found and sequenced [36], the psbA gene from an atrazine-tolerant *Amaranthus* was then sequenced [37]. Based on their findings, the resistance is due to a chloroplast genome mutation of the psbA gene, which codes the D1 protein. The molecular analysis showed that resistance is due to the substitution of serine 264 to glycine (Ser-264-Gly) in many weed species [38–42]. The substituted urea herbicides, as PS II inhibitors [43] also bind in a niche on the D1 protein, but not at the identical site as the triazines.

A schematic diagram of the folding of the herbicide-binding site on the D1 protein [44], updated with further amino acids in triazine resistance, is given in **Figure 2** [45]. From total of 345 amino acids in the D1 protein, around 60 are part of the herbicide and $Q_B$-binding site. Arrows indicate possible mutations (such as Val-219, Ala-251, Phe-255, Gly-256, Ser-264

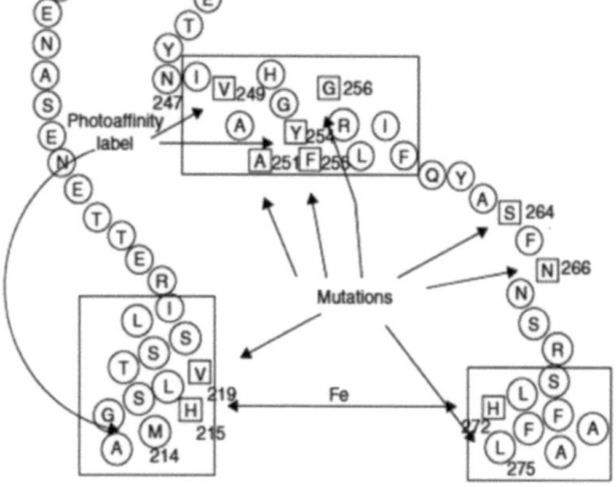


**Figure 2.** The amino acid sequence of the herbicide-binding D1 protein in the PS II. This work of Michel and Deisenhofer was honoured with the Nobel Prize. (downloaded from Ref. [45]).

| Herbicide groups according the site of action | Number of weed species | | | | | Total |
|---|---|---|---|---|---|---|
| | 1970–1979 | 1980–1989 | 1990–1999 | 2000–2009 | 2010–2016 | |
| Inhibitors PS II (C1/5) | 20 | 32 | 11 | 7 | 4 | 74 |
| Inhibitors PS II (C2/7) | 1 | 6 | 13 | 4 | 4 | 28 |
| Inhibitors PS II (C3/6) | / | / | 1 | 3 | / | 4 |
| Inhibitors ESPSP enzyme | / | / | 2 | 18 | 16 | 36 |
| Inhibitors AHAS enzyme | / | 11 | 62 | 53 | 33 | 159 |
| Inhibitors ACC-ase | / | 5 | 21 | 14 | 8 | 48 |
| Total | 21 | 54 | 110 | 99 | 65 | |

**Table 2.** The first confirmed cases of weed species that have developed resistance to different herbicides site of action according to decades.

and Leu-275) in herbicide-resistant plants and algae or amino acids tagged by herbicides azidoderivatives (Met-214 by azidoatrazine) [45].

Currently, resistance to herbicides that target photosynthesis at PS II has been documented in 74 weed species for triazines (C1/5 group), 28 in C2/7 and only 4 in C3/6 according to the data in the **Table 2** [7]. Except the usual amino acid substitution Ser-264-Gly in the D1 protein, reduced absorption, translocation and/or detoxification have been reported very often for resistance to triazines in many weed species (**Table 3**).

However, diverse chemical groups of herbicides PS II inhibitors (according to HRAC: C1—triazineas, triazinones, triazolinone, pyridazinones, phenyl-carbametes, uracils; C2—amides, ureas; C3—benzothiadiazinones, nitriles, phenyl pyridazines) bind to overlapping, but not identical sites on the D1 protein [43]. Several different amino acid substitutions that confer resistance to herbicide PS II inhibitors have been identified in or near the $Q_B$-binding niche such as: Ser-264-Thr in *Portulaca oleracea* [71], Ser-264-Gly and Val-219-Ile in *P. annua* and *K. scoparia* [64, 68, 70], Asn-266-Thr in *S. vulgaris* [73] as well as Ser-264-Gly, Ala-251-Val and Leu-218-Val in *C. album* [41, 59]. In addition, dependence of herbicides, interaction between herbicides, specific amino acid substitution, varying levels of cross or negative cross-resistance have been reported for different mutations in the D1 protein [64]. Resistance ratios for *P. oleracea* a Ser-264-Thr mutant were 8 and >6 for linuron and diuron, respectively; >800 for atrazine; and >20 for terbacil. Linuron resistant *P. oleracea* was negatively cross-resistant to pyridate and bentazon (0.75 and 0.5, respectively) [71].

| Weed species | Mechanisms | |
|---|---|---|
| | Amino acid substitution | Other mechanisms of resistance |
| *Abutilon theophrasti* Medic. | | Detoxification [45–49] |
| *Alopecurus myosuroides* Huds. | | Detoxification [50–52] |
| *Amaranthus tuberculatus* Moq. Sauer. | | Detoxification [53, 54] |
| *Amaranthus retroflexus* L. | | Detoxification [55] |
| *Amaranthus hybridus* L. | Ser-264-Gly [37] | |
| *Amaranthus powellii* S. Wats. | | Detoxification [56] |
| *Bromus tectorum* L. | | Detoxification [57] |
| *Brassica napus* L. | Ser-264-Gly [58] | |
| *Chenopodium album* L. | Ser-264-Gly, Ala-251-Val, Leu-218-Val [41, 59] | Detoxification [55, 56, 60, 61] |
| *Echinochloa crus-galli* (L.) P. Beauv. | | Reduced absorption and translocation, detoxification [60, 62, 63] |
| *Kochia scoparia* (L.) Schr. | Val-219-Ile [64] | |
| *Lolium rigidum* Gaudin. | | Detoxification [65–67] |
| *Poa annua* L. | Ser-264-Gly, Val-219-Ile [40, 68–70] | |
| *Portulaca oleracea* L. | Ser-264-Thr [71] | |
| *Solanum nigrum* L. | Ser-264-Gly [72] | |
| *Senecio vulgaris* L. | Asn-266-Thr [73] | |
| *Vulpia bromoides* (L.) S.F.Gray. | Ser-264-Gly [42] | |

**Table 3.** Confirmed mechanisms of resistance to herbicide PS II inhibitors in some weed species.

## 3. Weed resistance to herbicide ALS inhibitors

Herbicide inhibitors of acetoacetate synthase (ALS) and acetohydroxyacid synthase (AHAS) belong to several chemical classes: sulfonylurea (SU), triazolopyrimidines (TPs), pyrimidinyl(thio) benzoates, sulfonylaminocarbonyltriazolinones, imidazolinones (IMIs). The first commercial SU herbicide was chlorsulfuron, which was introduced by DuPont in 1982 for weed control in small grain crops. The SUs are highly active herbicides, effective at use rates as low as 2 g a.i. $ha^{-1}$ [74]. Almost simultaneously, researchers at American Cyanamid discovered a structurally distinct family of herbicides, the IMIs, which were also shown to inhibit the ALS enzyme [75]. Since then, three additional chemical classes of ALS inhibitors have been discovered. Those products provide both pre-emergent and post-emergent control of many serious monocot and dicot weed species in many crops.

ALS is the first enzyme in the branched-chain amino acid pathway, which catalyzes the first steps in amino acid biosynthesis such as valine, leucine and isoleucine [76]. The first case of resistance to ALS inhibitors (chlorsulfuron) was reported within 5 years after the introduction of SU herbicides, in 1987 in the United States [77]. Herbicide-resistant weed evolution is more common for ALS inhibitors compared to herbicides of other groups. Currently, 159 weed species have evolved resistance to ALS-inhibiting herbicides [7] according to decades that could be seen in **Table 2**. Weed resistance to ALS inhibitors is due to an alteration of the gene encoding the ALS enzyme. The positions in ALS from various sources (plant, yeast, bacteria) where mutations are known to confer resistance to one or more herbicides distributed across the α, β and γ domain of the protein (**Figure 3**) [78]. Weed species or genera with high incidence of target-site ALS resistance include *Amaranthus* spp., *K. scoparia* and *Papaver rhoeas*, among others. Studies have shown that mutations of eight amino acid residues are known to be involved in causing weed resistance: Ala-122, Pro-197, Ala-205, Asp-376, Arg-377, Trp-574, Ser-653 and Gly-654 (**Table 4**).

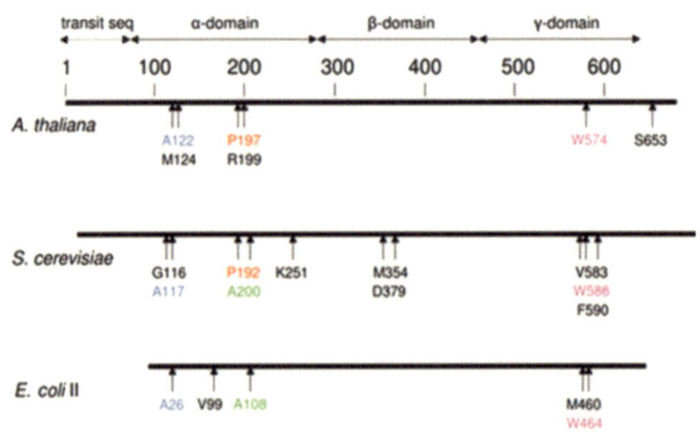


**Figure 3.** ALS mutations conferring herbicide resistance. Arrows point to positions in the sequences of ALS from different sources (plant, yeast, bacteria) where spontaneous or induced mutations result in an herbicide-insensitive enzyme. Colours designate substitutions occurring in more than one species (downloaded from Ref. [78]).

| Weed species | Mechanism of resistance | |
|---|---|---|
| | **Amino acid substitutions** | **Other mechanisms of resistance** |
| *Amaranthus retroflexus* L. | Ala-122-Thr, Pro-197-Leu, Ala-205-Val, Asp-376-Glu, Arg-377-His, Trp-574-Leu, Ser-653-Thr [80] | |
| *Amaranthus powellii* S. Warts. | Ala-122-Thr, Asp-376-Glu, Arg-377-His, Ser-653-Thr | |
| *Amaranthus hybridus* L. | Ala-122-Thr, Asp-376-Glu, Arg-377-His, Ser-653-Asn | Detoxification [86] |
| *Amaranthus blitoides* S. Wats | Pro-197-Ser, Arg-377-His | |
| *Amaranthus tuberculatus* (Moq.) Sauer | Arg-377-His, Ser-653-Asn/Thr | Altered enzyme activity, detoxification [87, 88] |
| *Amaranthus palmeri* (S.) Warts. | Arg-377-His, Ser-653-Asn | Altered enzyme activity [87, 89] |
| *Ambrosia artemisiifolia* L. | Arg-377-His | |

| Weed species | Mechanism of resistance | |
|---|---|---|
| | **Amino acid substitutions** | **Other mechanisms of resistance** |
| *Ambrosia trifida* L. | Arg-377-His | Reduced translocation, detoxification, sequestration [90] |
| *Alopecurus aequalis* Sobol. | Pro-197-Thr, Arg-377-His | |
| *Alopecurus myosuroides* Huds. | Pro-197-Thr, Arg-377-His | |
| *Anthemis cotula* L. | Pro-197-Ser/Thr/Leu/Gln | |
| *Apera spica-venti* (L.) P.B. | Ala-122-Val, Pro-197-Ser/Thr/Ala/Asn, Arg-377-His, Trp-574-Leu/Met | |
| *Avena fatua* L. | Ser-653-Asn/Thr | |
| *Bromus tectorum* L. | Pro-197-Ser | |
| *Capsella bursa-pastoris* (L.) Med. | Pro-197-Ser/Thr/Leu/His | |
| *Camelina microcarpa* Andrz. | Arg-377-His | |
| *Conyza canadensis* (L.) Cronq. | Pro-197-Ser/Ala, Ala-205-Val, Asp-376-Glu, Trp-574-Leu | |
| *Cyperus difformis* L. | Pro-197-Ser/Ala/His | |
| *Cyperus iria* L., *C. escculentus* L. | Trp-574-Leu | |
| *Descurainia sophia* (L.) Webb. | Pro-197-Ser/Thr/Leu/Ala/His/Tyr, Asp-376-Glu, Arg-377-Leu | |
| *Echinochloa crus-galli* L. | Ala-122-Thr/Val, Arg-377-His | Detoxification [91] |
| *Echinochloa phyllopogon* (Stapf) Koss | Arg-377-His | Detoxification [92] |
| *Galium aparine* L. | Trp-574-Gly | |
| *Galium spurium* L. | Asp-375-Glu, Trp-574-Leu, Ser-653-Asn | |
| *Helianthus annuus* L. | Pro-197-Leu, Ala-205-Val | Altered enzyme activity [90] |
| *Kochia scoparia* (L.) Schrad. | Pro-197-Ser/Thr/Leu/Ala/Gln/Arg, Asp-376-Glu, Trp-574-Leu | Altered enzyme activity [93] |
| *Lactuca serriola* L. | Pro-197-Thr/His | |
| *Lamium amplexicaule* L. | Pro-197-Arg | |
| *Lolium perenne* L. | Asp-376-Glu | |
| *Lolium rigidum* Gaud. | Pro-197-Ser/Leu/Ala/Gln/Arg, Trp-574-Leu | Detoxification altered enzyme activity [94, 95] |
| *Myosoton aquaticum* (L.) Moench. | Pro-197-Ser/Glu | Detoxification [96] |
| *Papaver rhoeas* L. | Pro-197-Ser/Thr/Leu/Ala/His/Arg, Trp-574-Leu | |
| *Poa annua* L. | Ala-205-Phe, Trp-574-Leu | |
| *Polygonum convolvulus* L. | Trp-574-Leu | |
| *Raphanus raphanistrum* L. | Ala-122-Try, Pro-197-Ser/Thr/Ala/His, Asp-376-Glu, Trp-574-Leu | |
| *Schoenoplectus juncoides* Roxb. | Pro-197-Ser/Leu/His, Asp-376-Glu, Trp-574-Leu | |

| Weed species | Mechanism of resistance | |
|---|---|---|
| | Amino acid substitutions | Other mechanisms of resistance |
| *Schoenoplectus mucronatus* (L.) Palla | Pro-197-His, Trp-574-Leu | |
| *Senecio vulgaris* L. | Pro-197-Ser/Leu | |
| *Setaria viridis* (L.) Beauv. | Ser-653-Asn/Thr/Ile, Gly-654-Asp | Altered enzyme activity [97] |
| *Sinapis arvensis* L. | Pro-197-Ser, Asp-376-Glu, Trp-574-Leu | |
| *Sisymbrium orientale* Torn. | Pro-197-Ile, Trp-574-Leu | |
| *Solanum ptycanthum* Dunn | Ala-122-Thr, Ala-205-Val | |
| *Sonchus asper* (L.) Mill. | Pro-197-Leu | |
| *Sorghum bicolor* (L.) Moench | | Altered enzyme activity [98, 99] |
| *Sorghum halepense* (L.) Pers. | Trp-574-Leu [100] | |
| *Stellaria media* (L.) Vill. | Pro-197-Gln, Trp-574-Leu | Altered enzyme activity [101] |
| *Thlaspi arvense* L. | Pro-197-Leu | |
| *Xanthium strumarium* L. | Ala-122-Thr, Ala-205-Val, Trp-574-Leu | Altered enzyme activity [102] |

Amino acid substitution in weed-resistant species to ALS inhibitors downloaded from HRAC [103].

**Table 4.** Confirmed mechanisms of resistance to herbicide ALS inhibitors in some weed species.

The most different amino acid substitutions in $\alpha$-domain at position Pro-197 have been linked in confirmed weed-resistant species such as: *K. scoparia* (Pro-197-Ser/Thr/Leu/Ala/Gln/Arg), *Descurainia sophia* (Pro-197-Ser/Thr/Leu/Ala/His/Tyr), *P. rhoeas* (Pro-197-Ser/Thr/Leu/Ala/His/Arg), *L. rigidum* (Pro-197-Ser/Leu/Ala/Gln/Arg), *Apera spica-venti* (Pro-197-Ser/Thr/Ala/Asn), etc. Also, the substitution of Trp-574-Leu confers resistance to several weed species (*A. retroflexus, C. iria, D. sophia, C. canadensis, K. scoparia, P. annua* etc.) and the levels of resistance are all high against SUs, IMIs and TPs (cross-resistance) [29, 79–83]. Generally, the low number of confirmed weeds resistant to ALS inhibitors is due to altered enzyme activity, reduced translocation and detoxification. Additionally, many weed populations resistant to ALS inhibitors have developed multiple-resistance to other chemical classes with different modes of action (e.g. auxinic herbicides, EPSPS inhibitors, ACC-ase inhibitors) [31, 84, 85].

## 4. Weed resistance to herbicides ACC-ase inhibitors

Herbicides acetyl-CoenzymeA carboxylase (ACC-ase) inhibitors are aryloxyphenoxy-propionates (APPs/FOPs), cyclohexanediones (CHDs/DIMs) and phenylpyrazoline. The first herbicide ACC-ase inhibitors commercialized in 1975 [104]. They are used as foliar

http://dx.doi.org/10.5772/67979

herbicides to control monocot weed species in dicot crops and some of them even in cereals or in rice. The mode of action of these herbicides is inhibition of fatty acid biosynthesis through blocking of the acetyl-CoenzymeA carboxylase [105]. Inhibition of lipid biosynthesis can explain the reduction of growth, increase in permeability of membrane and the ultrastructural effects commonly observed. In living organisms, ACC-ase exists in two different types: multi-subunit type and multi-functional type with 17–51 kDa (prokaryote) and 220–280 kDa (eukaryote) in size, respectively [106]. In dicot plants, the enzyme is structurally distinguished from the enzyme of monocots which contains four regions (biotin carboxylase, biotin carboxy carrier protein, carboxyl-transferase $\alpha$ and $\beta$), while in dicots, they are encoded on separate proteins.

The frequent use of FOPs and DIMs has resulted in the development of resistance to ACC-ase inhibitors in some monocot species in many countries in the world. Currently, 48 weed species have evolved resistance to these herbicides [7]. By decades, dynamics of the confirmation of the first cases of resistant weed species to the ACC-ase can be seen in **Table 2**. Generally, mechanisms of resistance to ACC-inhibiting herbicides can be divided in two categories: ACC-related and metabolism-based. Target-site resistance to ACC-ase inhibitors due to the herbicides binding to the carboxyl-transferase region within the ACC-ase enzyme results in amino acid substitution in that region (**Figure 4**) [107, 108]. Weed species or genera with high affinity of target-site ACC-ase resistance are *A. myosuroides*, *Avena* sp., *Bechmannia syzigachne*, *E. crus-galli*, *Lolium* sp., etc. Most commonly amino acid substitution such as Ile-1781-Leu, Trp-1999-Cys, Trp-2027-Cys, Ile-2041-Asn, Asp-2078-Gly, Cys-2088-Arg, Gly-2096-Ser was confirmed in monocot resistant populations of weed species [109–116]. Amino acid substitutions such as Asp-2078-Gly and Cys-2088-Arg usually provide strong level of resistance to all ACC-ase (FOPs, DIMs, pinoxaden) inhibitors [81]. Moreover, altered enzyme activity, gene expression and detoxification were very often included in weed resistance to ACC-ase inhibiting herbicides (**Table 5**). Also, in some population of weed species such as *A. myosuroides* [117], *E. crus-galli* [118], *L. rigidum* [111] and *Lolium perenne* [85], target and non-target multiple-resistance, which involves ACC-ase and ALS inhibitors or ACC-ase and EPSPS inhibitors, was confirmed.

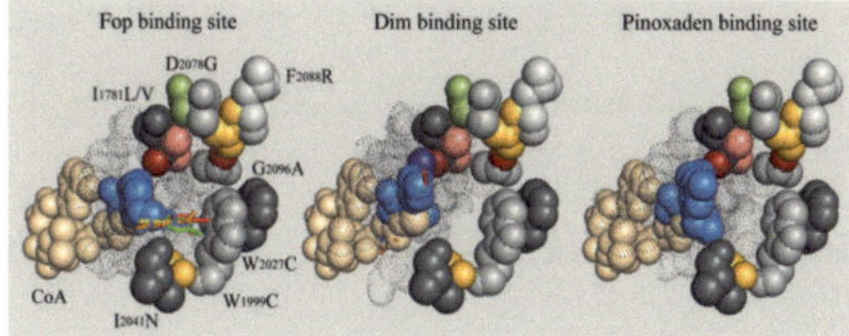


**Figure 4.** Single amino acid mutations in acetyl-CoA carboxylase in monocot-resistant weed populations (downloaded from Ref. [108]).

| Weed species | Mechanism of resistance | |
|---|---|---|
| | Amino acid substitutions | Other mechanisms of resistance |
| *Alopecurus aequalis* Sobol. | Ile-1781-Leu [119] | |
| *Alopecurus myosuroides* Huds. | Ile-1781-Leu, Trp-2027-Cys, Ile-2041-Asn, Asp-2078-Gly, Gly-2096-Ala [110, 112] | Detoxification, gene expression [120–122] |
| *Avena fatua* L. | Ile-1781-Leu, Trp-1999-Cys,Trp-2027-Cys, Ile-2041-Asn, Asp-2078-Gly, Cys-2088-Arg, Gly-2096-Ser [109, 114] | Detoxification [123, 124] |
| *Avena sterilis* L. | Ile-1781-Leu, Trp-1999-Cys, Trp-2027-Cys, Ile-2041-Asn, Asp-2078-Gly, Cys-2088-Arg [115] | Detoxification [123, 125] |
| *Bechmannia syzigachne* (Steud.) Fernald | Ile-1781-Leu, Ile-2041-Asn, Asp-2078-Gly [126, 127] | |
| *Echinochloa crus-galli* (L.) Beauv. | Ile-1781-Leu [128] | Altered enzyme activity, gene expression [129] |
| *Echinochloa colona* (L.) Link. | | altered enzyme activity [130] |
| *Eleusine indica* (L.) Gaertn | Asp-2078-Gly, Thr-1805-Ser [131] | |
| *Hordeum glaucum* (Steud.) Tzvelev | Ile-1781-Leu, Gly-2096-Ala [132] | |
| *Hordeum leporinum* (Link) Arcang. | Ile-1781-Leu, Gly-2096-Ala [132] | Detoxification, altered enzyme activity [133] |
| *Lolium multiflorum* Lam. | Ile-1781-Leu (Ile-418-Leu), Cys-2088-Arg [134, 135] | Detoxification [136] |
| *Lolium perenne* L. ssp. *multiflorum* Lam. | Ile-1781-Leu, Trp-2027-Cys, Ile-2041-Asn, Asp-2078-Gly [137, 138] | |
| *Lolium rigidum* Gaud. | Ile-1781-Leu, Ile-2041-Asn, Asp-2078-Gly, Cys-2088-Arg, Gly-2096-Ala, Trp-2027-Cys [111, 116] | Detoxification [139, 140] |
| *Lolium* sp. | Ile-1781-Leu, Trp-1999-Cys, Ile-2041-Asn/Val, Asp-2078-Gly, Cys-2088-Arg, Gly-2096-Ala [113] | |
| *Pseudosclerochloa kengiana* | Trp-1999-Ser [141] | |
| *Setaria viridis* L. Beauv. | Ile-1780-Leu [142] | Altered enzyme activity [143] |
| *Sorghum halepense* (L.) Pers. | Ile-2041-Asn [144] | Altered enzyme activity [145] |

**Table 5.** Confirmed mechanisms of resistance to herbicide ACC-ase inhibitors in some weed species.

## 5. Weed resistance to herbicide EPSPS inhibitors, glyphosate

Glyphosate was discovered and developed as a non-selective herbicide by Chemical Company Monsanto in 1974. N-(phosphonometil) glycine, the active ingredient in glyphosate, is a derivate of the amino acid glycine and phosphonic acid. It's mode of action in relation to the enzyme EPSPS (5-enolpyruvylshikimate-3-phosphate synthase) preventing the biosynthesis of the aromatic amino acids required for the production of growth regulators, anthocyanins, phenolics and proteins [146]. The site of action is located in the chloroplast and it was confirmed 8 years after glyphosate introduction [146]. Broad weed spectrum (annual and perennial, monocots and dicots), high efficacy, lack of soil activity and low mammalian toxicity are key characteristics that make glyphosate the world's most widely used herbicide [147]. Because glyphosate is inherently non-selective, selectivity has often been achieved by placement and timing, for example, as a pre-plant or pre-emergence herbicide for the control of weeds in no-till systems and for turf-grass renovation [104]. The introduction of genetically modified glyphosate resistant crops in the United States and other parts of the world [148] has led to enormous increase of glyphosate use on arable land (cotton, canola, corn, wheat, sugar beets, potatoes, etc.) as a post-emergence herbicide.

Glyphosate-resistant weeds were not found during the first 15 years of glyphosate use (1972–1997). Based on the resistance risk criteria for assessing the risk of developing weed resistance to glyphosate, it was estimated that the glyphosate has low risk for the evolution of weed resistance [149]. However, in the last 19 years (1998–2016), glyphosate resistance in 36 weed species was confirmed and according to the decades, it looks like this: 2 (first decade), 18 (second) and 16 species (the last, third) (**Table 2**) [7]. Mechanism of glyphosate resistance to weed species includes target-site mutation, target-site gene amplification/expression, active vacuole sequestration, limited cellular uptake and a rapid necrosis response [21].

In a number of cases of confirmed weed resistance to glyphosate, the resistance was based on some different mechanisms which include non-target-site (limited absorption and translocation, vacuolar sequestration) and target-site resistance (amino acid substitution, ESPSP gene expression/amplification, altered enzyme activity) (**Table 6**). Generally, usually confirmed

| Weed species | Mechanisms | |
|---|---|---|
| | Amino acid substitution | Other mechanisms of resistance |
| *Amaranthus tuberculatus* Moq. Sauer. | Pro-106-Ser [150, 151] | Gene expression [150, 151] |
| *Amaranthus palmeri* S. Wats. | | Gene expression [23, 24, 152] |
| *Abutilon theophrasti* Medic. | | Reduced absorption and translocation [153, 154] |
| *Cirsium arvense* (L.) Scop. | | Reduced absorption and translocation [155] |
| *Conyza canadensis* (L.) Cronq. | | Reduced absorption and translocation, detoxification, vacuole sequestration [26, 156–160] |
| *Conyza bonariensis* (L.) Cronq. | | Reduced absorption and translocation [161] |
| *Chenopodium album* L. | | Reduced absorption and translocation [162] |
| *Cyperus esculentus* L. | | Reduced absorption and translocation [163] |
| *Cyperus rotundus* L. | | Detoxification [164] |
| *Eleusina indica* (L.) Gaertn. | Pro-106-Ser/Thr [165–167] | |

| Weed species | Mechanisms | |
|---|---|---|
| | Amino acid substitution | Other mechanisms of resistance |
| *Echinochloa colona* (L.) Link. | Pro-106-Ser [168] | |
| *Kochia scoparia* (L.) Schr. | | Detoxification, gene expression [25] |
| *Lolium rigidum* Gaudin. | Pro-106-Ser/Ala/Thr/Leu [169, 170] | Reduced absorption and translocation, altered enzyme activity [171–177] |
| *Lolium multiflorum* Lam. | Pro-106-Ser [178] | Reduced absorption and translocation [178, 179] |
| *Lolium perenne* L. ssp. *multiflorum* Lam. | | Gene expression [180] |
| *Poa annua* L. | Pro-106-Ala [181] | Reduced translocation [181] |
| *Sorghum halepense* (L.) Pers. | | Reduced absorption and translocation [182] |

**Table 6.** Confirmed mechanisms of resistance to EPSPS inhibitor in some weed species.

cases of weed resistance to glyphosate were due to reduced absorption and translocation of the herbicide. Further, cDNA sequence analysis of the EPSPS gene indicated that resistance to glyphosate was based on substitution of proline with serine (Pro-106-Ser), alanine (Pro-106-Ala), threonine (Pro-106-Thr), or leucine (Pro-106-Leu)) at the position 106 of the EPSPS protein in many weed species (*Amaranthus tuberculatus, E. indica, E. colona, L. rigidum, Lolium multiflorum, P. annua*).

## 6. Management strategies for herbicide-resistant weeds

Strategy for herbicide-resistance weed management must involve all the available preventive, cultural, mechanical and chemical measures for effective, safe and cost-effective weed control [183]: (a) survey of present weed flora; (b) preventing weed seed production and reduction of weed seed in the soil seed-bank; (c) prevention of the movement of seeds and vegetative propagules from field to field or from field margins (or lost field) to field; (d) keep arable and non-arable land as weed free as possible; (e) sowing pure crop seeds; (f) growing competitive crops that can suppress weeds; (g) destruction of weed seeds in post-harvest materials (e.g. Integrated Harrington Seed Destructor); (h) use mechanical and physical measures where appropriate; (i) using herbicides with different modes of action, tank mixtures and sequential applications; (j) use of recommended herbicide rate for certain number of weed populations; (k) adopting crop rotations that allow use of herbicides of alternative mode of action; (l) intensify research and professional communication and grower education programs and (m) publish guidelines for managing anti-resistant strategy.

http://dx.doi.org/10.5772/67979

The state government sectors, universities and research institutes, technology development centres, farmers and other relevant stakeholders were called to proactively address emerging weed resistance problems and to develop cost-effective resistance-management strategy and practices that support effective weed control.

## Acknowledgements

This work was supported by the Ministry of Education, Science and Technological Development of Republic of Serbia.

## Author details

Sava Vrbničanin[1]*, Danijela Pavlović[2] and Dragana Božić[1]

*Address all correspondence to: sava@agrif.bg.ac.rs

1 Faculty of Agriculture, University of Belgrade, Belgrade, Serbia

2 Institute for Plant Protection and Environment, Belgrade, Serbia

## References

[1] LeBaron HM, Gressel J. Herbicide Resistance in Plants. Wiley Interscience NYP; 1982

[2] Penner D. Herbicide action and metabolism. In: Turf Weeds and their Control. Madison, WI: American Society of Agronomy and Crop Science Society of America; 1994. pp. 37-70

[3] Gressel J. Why get resistance? It can be prevented or delayed. In: Caseley JC, Cussans GW, Atkin RK, editors. Herbicide Resistance in Weeds and Crops. Butterworth-Heinemann Ltd; 1991. pp. 1-27

[4] Preston C, Mallory-Smith CA. Biochemical mechanisms, inheritance, and molecular genetics of herbicide resistance in weeds. In: Powles SB, Shanel DL, editors. Herbicide Resistance in World Grains. Boca Raton, FL: CRC Press; 2001. pp. 23-61

[5] Heap I. Herbicide resistant weeds. In: Primentel D, Peshin R, editors. Integrated Pest Management. Pesticide Problems. Vol. 3. Springer; 2014. pp. 281-303

[6] Ryan GF. Resistance of common groundsel to simazine and atrazine. Weed Science. 1970;**18**:614-616

[7] Heap I. The International Survey of Herbicide Resistant Weeds [Internet]. Available from: http://www.weedscience.org [Accessed: 3 January 2017]

[8] Janjić V, Stanković-Kalezić R, Radivojević Lj, Marisavljević D, Jovanović Lj, Ajder S. Resistance of *Amaranthus retroflexus* L. and *Chenopodium alabum* L. to atrazine (In Serbian). Acta biologica Iugoslavica, Series G: Acta Herbologica. 1994;**3**(1):63-73

[9] Janjić V, Stanković-Kalezić R, Radivojević Lj, Ajder S, Marisavljević D, Jovanović LJ. Distribution of atrazine resistant population of *Amaranthus retroflexus* L. (In Serbian). Acta Biologica Iugoslavica, Series G: Acta Herbologica. 1994;**3**(1):23-31

[10] Janjić V, Stanković-Kalezić R, Radivojević Lj, Marisavljević D, Ajder S, Jovanović LJ. Detection some triazine resistant weeds using chlorophyll fluorescence. In: Proceedings of the International Symposium on Weed and Crop Resistance to Herbicides. Cordoba, Spain: 1995. p. 134

[11] Malidža G, Rajković M, Vrbničanin S, Božić D. Cross-resistance of *Sorghum halepense* to ALS inhibitors in Serbia and implications for resistance management. Book of Abstracts of the VII Congress on Plant Protection: Integrated Plant Protection Knowledge—Based Step Towards Sustainable Agriculture, Forestry and Landscape Architecture. 2014:143-144

[12] Malidža G, Rajković M, Vrbničanin S, Božić D. Identification and distribution of ALS resistant *Sorghum halepense* populations in Serbia. Book of Abstracts of the 17th European Weed Research Society Symposium; 2015:115

[13] Pavlović D, Vrbničanin S, Božić D, Fischer A. Morphophysiological traits and triazine sensitivity in *Chenopodium album* L. Pest Management Science. 2008;**64**(2):101-107

[14] Božić D, Vrbničanin S, Barać M, Stefanović L. Determination of Jonsongrass (*Sorghum halepense* (L.) Pers.) level of sensitivity to nicosulfuron. Maydica. 2007;**52**(3):271-277

[15] Pavlovic D, Vrbnicanin S, Reinhardt C. Crop response to glyphosate trimesium sulphosate. Notulae Botanicae Horti Agrobotanici Cluj-Napoca. 2013;**41**(2):582-589

[16] Bozic D, Barac M, Saric-Krsmanovic M, Pavlovic D, Ritz C, Vrbnicanin V. Common cocklebur (*Xanthium strumarium*) response to nicosulfuron. Notulae Botanicae Horti Agrobotanici Cluj-Napoca. 2015;**43**(1):186-191

[17] Pavlović D, Vrbničanin S, Reinhard C, Marisavljević D. Morpho-anatomical response of glyphosate-resistant and –susceptible maize to glyphosate trimesium. In: Proceedings of the International Symposium on Current Trends in Plant Protection. Belgrade, Serbia; 2012. pp. 188-191

[18] Pavlović D, Vrbničanin S, Elezović I, Marisavljević D, Božić D. Physiological parameters as a basis for differentiating between resistant and susceptible populations of *Chenopodium album* L. to atrazine (in Serbian). Pesticidi i fitomedicina. 2005;**20**(4):241-247

[19] Bozic D, Pavlovic D, Bregola V, Di Loreto A, Bosi S, Vrbnicanin S. Gene flow from Herbicide-Resistant sunflower hybrids to weedy sunflower. Journal of Plant Diseases and Protection. 2015;**122**(4):183-188

[20] Heap I, LeBaron H. Introduction and overview of resistance. In: Powles SB, Shanel DL, editors. Herbicide Resistance in World Grains. Boca Raton, Florida: CRC Press; 2001. pp. 1-22

[21] Sammons RD, Gaines TA. Glyphosate resistance: State of knowledge. Pest Management Science. 2014;**70**:1367-1377

[22] Yu Q, Powles S. Metabolism-based herbicide resistance and Cross-Resistance in crop weeds: A threat to herbicide sustainability and global crop production. Plant Physiology. 2014;**166**:1106-1118

[23] Gaines TA, Zhang W, Wang D, Bukun B, Chisholm ST, Shaner DL, Nissen SJ, Patzoldt WL, Tranel PJ, Culpepper AS, Grey TL, Webster TM, Vencill WK, Sammons RD, Jiang J, Preston C, Leach JE, Westra P. Gene amplification confers glyphosate resistance in *Amaranthus palmeri*. Proceedings of the National Academy of Sciences. 2010;**107**(3):1029-1034

[24] Vila-Aiub MM, Goh SS, Gaines TA, Han H, Busi R, Yu Q, Powles SB. No fitness cost of glyphosate resistance endowed by massive EPSPS gene amplification in *Amaranthus palmeri*. Planta. 2014;**239**(4):793-801

[25] Godar AS, Stahlman PW, Jugulam M, Dille JA. Glyphosate-Resistant Kochia (*Kochia scoparia*) in Kansas: EPSPS gene copy number in relation to resistance levels. Weed Science. 2015;**63**:587-595

[26] Ge X, D'Avignon DA, Ackermana JJH, Sammons RD. Rapid vacuolar sequestration: The horseweed glyphosate resistance mechanism. Pest Management Science. 2010;**66**: 345-348

[27] Ge X, D'Avignon DA, Ackerman JJH, Collavo A, Sattin M, Ostrander EL. Hall EL, Sammons RD, Preston C. Vacuolar glyphosate-sequestration correlates with glyphosate resistance in ryegrass (*Lolium* spp.) from Australia, South America, and Europe: A 31P NMR investigation. Journal of Agriculture and Food Chemistry. 2012;**60**:1243-1250

[28] Dekker JH, Duke SO. Herbicide-Resistant Field Crops [Internet]. Iowa State University, Digital Repository. 1995. Available from: http://lib.dr.iastate.edu/agron_pubs (Accessed: 5 December 2016)

[29] Riar DS, Tehranchian P, Norsworthy JK, Nandula V, McElroy S, Srivastava V, Chen S, Bond JA, Scott RC. Acetolactate synthase-inhibiting, herbicide-resistant rice flatsedge (*Cyperus iria*): Cross-resistance and molecular mechanism of resistance. Weed Science. 2015;**63**:748-757

[30] LeBaron HM, Utah H. Distribution and management of Triazine-Resistant weeds. In: LeBaron HM, McFarland EJ, Burnside CO, editors. The Triazine Herbicides 50 years Revolutionizing Agriculture. Elsevier; 2008. pp. 119-133

[31] Collavo A, Sattin M. First glyphosate-resistant *Lolium* spp. biotypes found in a European annual arable cropping system also affected by ACCase and ALS resistance. Weed Science. 2014;**54**:325-334

[32] Délye C, Jasieniuk M, Le Corre V. Deciphering the evolution of herbicide resistance in weeds. Trends in Genetics. 2013;**29**(11):649-658

[33] Müller G. History of the discovery and development of triazine herbicides 13. In: LeBaron HM, McFarland EJ, Burnside CO, editors. The Triazine Herbicides 50 years Revolutionizing Agriculture. Elsevier; 2008. pp. 13-31

[34] Good NE. Inhibitors of the hill reaction. Plant Physiology. 1961;**36**:788-803

[35] Trebst A. The mode of action of triazine herbicides in plants. In: LeBaron HM, McFarland EJ, Burnside CO, editors. The Triazine Herbicides 50 years Revolutionizing Agriculture. Elsevier; 2008. pp. 101-111

[36] Zurawski G, Bohnert HJ, Whitfield PR, Bottomley W. Nucleotide sequence of the gene for the Mr 32 000 thylakoid membrane protein from *Spinacia oleracea***and** *Nicotiana debneyi*predicts a totally conserved primary translation product of Mr 38,950. Proceedings of the National Academy of Sciences United State of America. 1982**;79:**7699-7703

[37] Hirschberg J, McIntosh L. Molecular basis of herbicide resistance in *Amaranthus hybridus*. Science. 1983;**222**(4630):1346-1349

[38] Morden CW, Golden SS. A *psb*A genes indicate common ancestry of prochlorophytes and chloroplasts. Nature. 1989;**337**:382-385

[39] Van Oorschot JC. Chloroplastic resistance of weeds to triazines in Europe. In: Caseley JC, Cussans WG, Atkin KR, editors. Herbicide Resistance in Weeds and Crops. Boston: Butterworth-Heinemann; 1991. pp. 87-102

[40] Kelly ST, Coats GE, Luthe DS. Mode of resistance of Triazine-Resistant annual bluegrass (*Poa annua*). Weed Technology. 1999;**13**:747-752

[41] Mechant E, De Marez T, Hermann O, Bulcke R. Resistance of *Chenopodium album* **to** photosystem II-inhibitors. Communications in Agricultural and Applied Biological Sciences. 2008**;73(4):**913-917

[42] Ashworth MB, Han H, Knell G, Powles SB. Identification of triazine-resistant *Vulpia bromoides*. Weed Technology. 2016;**30**:456-463

[43] Gronwald JW. Resistance to photosystem II inhibiting herbicides. In: Powles SB, Holtum JAM, editors. Herbicide Resistance in Plants: Biology and Biochemistry. Boca Raton, Florida: Lewis Publishers; 1994. pp. 27-60

[44] Trebst A. The three dimensional structure of the herbicide binding niche on the reaction center polypeptide of photosystem II. Z. Naturforschg. 1987;**42c**:742-750

[45] Gronwald JW, Anderson RN, Yee C. Atrazine resistance in velvetleaf (*Abutilon theophrasti*) due to enhanced atrazine detoxification. Pesticide Biochemistry and Physiology. 1989;**34**:149-163

[46] Stowe AE, DiTomaso JM. Evidence for increased herbicide detoxification in triazine tolerant velvetleaf. In Proceedings of the 43rd Annual Meeting of the NorthEastern; Weed Science Society; 1989. pp. 31-35

[47] Anderson MP, Gronwald JW. Atrazine resistance in a velvetleaf (*Abutilon theophrasti*) biotype due to enhanced glutathione S-Transferase activity. Plant Physiology. 1991;**96**: 104-109

[48] Plaisance KL, Anderson MP, Gronwald JW. Glutathione S-transferase isoforms in atrazine-resistant and susceptible biotypes of *Abutilon theophrasti*. In: Proceedings of the International Symposium on Weed and Crop Resistance to Herbicides, Cordoba Spain; 1995. p. 99

[49] Pavlovic D, Vrbnicanin S, Bozic D, Simoncic A. *Abutilon theophrasti* Medic. Population responses to atrazine. Journal Central European Agriculture. 2007;**8**(4):435-442

[50] Hall LM, Moss SR, Powles SB. Mechanism of resistance to chlorotoluron in two biotypes of the grass weed *Alopecurus myosuroides*. Pesticide Biochemistry and Physiology. 1995;**53**:180-192

[51] Reade HPJ, Belfield LJ, Cobb HA. Rapid tests for herbicide resistance in black-grass based on elevated gluthatione S-transferase activity and abundance. Proceedings of the Brighton crop Protection Conference-Weeds. 1999;184-190

[52] Milner LJ, Reade PHJ, Cobb HA. An investigation of gluthatione S-transferase activity in *Alopecurus myosuroides* Huds. (black-grass) in the field. Proceedings of the Brighton crop Protection Conference-Weeds. 1999;173-178

[53] Patzoldt WL, Dixon BS, Tranel PJ. Triazine resistance in *Amaranthus tuberculatus* (Moq) Sauer. that is not site-of-action mediated. Pest Management Science. 2003;**59**:1134-1142

[54] Huffman J, Hausman EN, Hager GA, Riechers ED, Tranel JP. Genetics and inheritance of nontarget-site resistances to atrazine and mesotrione in a Waterhemp (*Amaranthus tuberculatus*) population from Illinois. Weed Science. 2015;**63**:799-809

[55] Pavlovic D, Vrbnicanin S, Elezovic I, Jovanovic LJ, Marisavljevic D. Alterations in amount of chlorophyll as indicator of resistance for *Chenopodium album* L. and *Amaranthus retroflexus* L. to atrazine. Journal of Plant Disease and Protection. 2006;131-138

[56] Khan US, Warwick IS, Marriage BP. Atrazin metabolism in resistant and susceptible biotypes of *Chenopodium album* L., *Chenopodium strictum* Roth. and *Amaranthus powelli* S. wats. Weed Research. 1985;**25**:33-37

[57] Menendez J, Bastida F, De Prado R. Resistance to chlortoluron in a downy brome (*Bromus tectorum*) biotype. Weed Science. 2006;**54**:237-245

[58] Sundby C, Chow WS, Anderson JM. Effects on photosystem II function, photoinhibition and plant performance of the spontaneous mutation of serine 264 in the photosystem II reaction center $D_1$ protein in triazine-resistant *Brassica napus* L. Plant Physiology. 1993;**103**(1):105-113

[59] Thiel H, Varrelmann M. Identification of a new PSII target site psbA mutation leading to D1 amino acid Leu218Val exchange in the *Chenopodium album* D1 protein and comparison to cross-resistance profiles of known modifications at positions 251 and 264. Pest Management Science. 2014;**70**(2):278-285

[60] Giebel J, Stachecki S, Praczyk T. The level of polyamines as an indicator of resistance or susceptibility of *Chenopodium album* to atrazine. Proceedings of the Brighton Crop Protection Conference-Weeds. 1999;163-166

[61] Giebel J, Praczyk T. Modification of IAA-oxidase activity by factor present in leaves of *Chenopodium album* and *Echinochloa crus-galli* in relation to susceptible-resistant plant response to triazine. Journal of Plant Protection Research. 2002;**42**(3):271-278

[62] Carey III FV, Duke OS, Hoaland ER, Talbert ER. Resistance mechanism of propanil resistant Barnyard grass I: Adsorption, translocation and site of action studies. Pesticide Biochemistry and Physiology. 1995;**52**:182-189

[63] Hoagland ER, Norsworty KJ, Carey F, Talbert ER. Metabolically based resistance to the herbicide propanil in *Echinochloa* species. Weed Science. 2004;**52**:475-486

[64] Mengistu LW, Christoffers MJ, Lym RG. A *psb*A mutation in *Kochia scoparia* (L) Schrad from railroad rights-of-way with resistance to diuron, tebuthiuron and metribuzin. Pest Management Science. 2005;**61**:1035-1042

[65] Burnet MWM, Loveys BR, Holtum JAM, Powles SB. A mechanism of chlorotoluron resistance in *Lolium rigidum*. Planta 1993;**190**:182-189

[66] Burnet MWM, Loveys BR, Holtum JAM, Powles SB. Increased detoxification is a mechanism of simazine resistance in *Lolium rigidum*. Pesticide Biochemistry and Physiology. 1993;**46**:207-218

[67] De Prado R, De Prado JL, Menendez J. Resistance to substituted urea herbicides in *Lolium rigidum* biotypes. Pesticide Biochemistry and Physiology. 1997;**57**:126-136

[68] Mengistu LW, Mueller, GW, Liston A, Barker RE. A *psb*A Mutation (valine219 to isoleucine) in *Poa annua* resistant to metribuzin and diuron. Pest Management Science. 2000;**56**:209-217

[69] Perry DH, McElroy JS, Dane F, van Santen E, Walker RH. Triazine-resistant annual bluegrass (*Poa annua*) populations with Ser 264 mutation are resistant to amicarbazone. Weed Science. 2012;**60**:355-359

[70] Svyantek AW, Aldahir P, Chen S, Flessner ML, McCullough PE, Sidhu SS, McElroy JS. Target and nontarget resistance mechanisms induce annual bluegrass (*Poa annua*) resistance to atrazine, amicarbazone, and diuron. Weed Technology. 2016;**30**(3):773-782

[71] Masabni JG, Zandstra BH. A serine-to-threonine mutation in linuron-resistant *Portulaca oleracea*. Weed Science. 1999;**47**:393-400

[72] Stankiewicz M, Gadamski G, Gawronski WS. Genetic variation and phylogenetic relationships of triazine-resistant and triazine-susceptible biotypes of *Solanum nigrum*—analysis using RAPD markers. Weed Research. 2001;**41**:287-300

[73] Park KW, Mallory-Smith CA. A *psb*A mutation (Asn266 to Thr) in *Senecio vulgaris* L. confers resistance to several PS II-inhibiting herbicides. Pest Management Science. 2006;**62**(9):880-885

[74] Beyer EM, Duffy MJ, Hay JV, Schlueter DD. Sulfonylureas. In: Kearney PC, Kaufman DD, editors. Herbicides: Chemistry, Degradation and Mode of Action. New York: Dekker; 1988. pp. 117-189

[75] Shaner DL, Anderson PC, Stidham MA. Imidazolinones: Potent inhibitors of acetohydroxyacid synthase. Plant Physiology. 1984;**76**:545-546

[76] Devine MD, Duke SO, Fedtke C. Physiology of Herbicide Action. Englewood Cliffs, NJ: Prentice Hall; 1993

[77] Shaner DL. Herbicide resistance: Where are we? How did we get here? Where are we going? Weed Technology. 1992;**9**:850-856

[78] Pang SS, Duggleby RG, Schowen RL, Guddat LW. The crystal structure of *Klebsiella pneumoniae* acetolactate synthase with enzyme-bound cofactor and with an unusual intermediate. The Journal of Biological Chemistry. 2004;**279**:2242-2253

[79] Thomson MK. Acetohydroxyacid synthase inhibitors (AHAS/ALS). In: Krämer W, Schirmer U, editors. Modern Crop Protection Compounds. Wiley-VCH; 2007. pp. 27-45

[80] Scarabel L, Varotto S, Sattin M. European biotype of *Amaranthus retroflexus* cross-resistant to ALS inhibitors and response to alternative herbicides. Weed Research. 2007;**47**:527-533

[81] Beckie HJ, Tardif FJ. Herbicide cross resistance in weeds. Crop Protection. 2012;**35:15-28**

[82] McElroy JS, Flessner ML, Wang Z, Dane F, Walker RH, Wehtje GR. A Trp574 to Leu amino acid substitution in the ALS gene of annual bluegrass (*Poa annua*) is associated with resistance to ALS-Inhibiting herbicides. Weed Science. 2013;**61**:21-25

[83] Deng W, Yanga Q, Zhang Y, Jiao H, Mei Y, Li X, Zheng M. Cross-resistance patterns to acetolactate synthase (ALS)-inhibiting herbicides of flixweed (*Descurainia sophia* L.) conferred by different combinations of ALS isozymes with a Pro-197-Thr mutation or a novel Trp-574-Leu mutation. Pesticide Biochemistry and Physiology. 2017;**136**:41-45

[84] Preston C, Dolman FC, Boutsalis P. Multiple resistance to acetohydroxyacid synthase–inhibiting and auxinic herbicides in a population of oriental mustard (*Sisymbrium orientale*). Weed Science. 2013;**61**:185-192

[85] Liu M, Hulting AG, Mallory-Smith C. Characterization of multiple herbicide-resistant Italian ryegrass (*Lolium perenne* ssp. *multiflorum*) populations from winter wheat fields in Oregon. Weed Science. 2016;**64**:331-338

[86] Manley BS, Wilson HP, Hines TH. Characterization of imidazolinone-resistant smooth pigweed (*Amaranthus hybridus*). Weed Technology. 1998;**12**(4):575-584

[87] Sprague CL, Stoller EW, Wax LM, Horak MJ. Palmer amaranth (*Amaranthus palmeri*) and common waterhemp (*Amaranthus rudis*) resistance to selected ALS-inhibiting herbicides. Weed Science. 1997;**45**(2):192-197

[88] Guo J, Riggins CW, Hausman NE, Hager AG, Riechers DE, Davis AS, Tranel PJ. Nontarget-Site resistance to ALS inhibitors in waterhemp (*Amaranthus tuberculatus*). Weed Science. 2015;**63**:399-407

[89] Burgos NR, Kuk YI, Talbert RE. *Amaranthus palmeri* resistance and differential tolerance to *Amaranthus palmeri* and *Amaranthus hybridus* to ALS-inhibitor herbicides. Pest Management Science. 2001;**57**:449-457

[90] Zelaya IA, Owen MDK. Evolved resistance to acetolactate synthase-inhibiting herbicides in common sunflower (*Helianthus annuus*), giant ragweed (*Ambrosia trifida*) and shattercane (*Sorghum bicolor*) in Iowa. Weed Science. 2004;**52**(4):538-548

[91] Riar DS, Norsworthy JK, Bond JA, Bararpour MT, Wilson MJ, Scott RC. Resistance of *Echinochloa crus-galli* populations to acetolactate synthase-inhibiting herbicides. International Journal of Agronomy. 2012;893-953

[92] Fischer AJ, Bayer DE, Carriere MD, Ateh CM, Yim KO. Mechanisms of resistance to Bispyribac-Sodium in an *Echinochloa phyllopogon* accession. Pesticide Biochemistry and Physiology. 2000;**68**:156-165

[93] Saari L, Cotterman J, Primiani M. Mechanism of sulfonylurea herbicide resistance in the broadleaf weed, *Kochia scoparia*. Plant Physiology. 1990;**93**:55-61

[94] Shimabukuro RH, Hoffer BL. Metabolism of diclofop-methyl in susceptible and resistant biotypes of *Lolium rigidum*. Pesticide Biochemistry and Physiology. 1991;**39**:251-260

[95] Christopher JT, Powles SB, Joseph AM, Holtum JAM. Resistance to acetolactate Synthase-Inhibiting herbicides in annual ryegrass (*Lolium rigidum*) involves at least two mechanisms. Plant Physiology. 1992;**100**:1909-1913

[96] Liu W, Wu C, Guo W, Du L, Yuan G, Wang J. Resistance mechanisms to an acetolactate synthase (ALS) inhibitor in water starwort (*Myosoton aquaticum*) populations from China. Weed Science. 2015;**63**:770-780

[97] Volenberg DS, Stoltenberg DE, Boerboom CM. Green foxtail (*Setaria viridis*) resistance to acetolactate synthase inhibitors. Phytoprotection. 2002;**83**:99-109

[98] Anderson DD, Nissen SJ, Martin AR, Roeth FW. Mechanism of primisulfuron resistance in a shattercane (*Sorghum bicolor*) biotype. Weed Science. 1998;**46**(2):158-162

[99] Lee DC, Martin RA, Roeth WF, Johnson EB, Lee JD. Comparison of ALS inhibitors resistance and allelic interactions in shattercane accessions. Weed Science. 1999;**47**(3):275-281

[100] Hernández MJ, León R, Fischer AJ, Gebauer M, Galdames R, Figueroa R. Target-site resistance to nicosulfuron in johnsongrass (*Sorghum halepense*) from Chilean corn fields. Weed Science. 2015;**63**:631-640

[101] Devine MD, Marles MAS, Hall LM. Inhibition of acetolactate synthase in susceptible and resistant biotypes of *Stellaria media*. Pesticide Science. 1991;**31**:273-280

[102] Lee JM, Owen MDK. Comparison of acetolactate synthase enzyme inhibition among resistant and susceptible *Xanthium strumarium* biotypes. Weed Science. 2000;**48**(3):286-290

[103] Tranel PJ, Wright TR, Heap IM. Mutations in Herbicide-resistant Weeds to ALS Inhibitors [Internet]. Available from: http://www.weedscience.com [Accessed: 10 January 2017]

[104] Monaco TJ, Weller SC, Ashton FM. Weed Science–Principles and Practices. 4th ed. John Wiley and Sons, Inc.; 2002

[105] Gronwald JW, Eberlein CV, Betts KJ, Baerg RJ, Ehlke NJ, Wyse DL. Mechanism of diclofop resistance in an Italian ryegrass (*Lolium multiflorum* Lam.) biotype. Pesticide Biochemistry and Physiology. 1992;**44**:126-139

[106] Kukorelli G, Reisinger P, Pinke G. ACCase inhibitor herbicides—selectivity, weed resistance and fitness cost: A review. International Journal of Pest Management. 2013;**59**(3):165-173

[107] Zhang H, Tweel B, Tong L. Molecular basis for the inhibition of the carboxyltransferase domain of acetylcoenzyme-A carboxylase by haloxyfop and dielofop. Proceedings of the National Academy of Sciences United State of America. 2004;**101**:5910-5915

[108] Jang SR, Marjanovic J, Gornicki P. Resistance to herbicides caused by single amino acid mutations in acetyl-CoA carboxylase in resistant populations of grassy weeds. New Phytologist. 2013;**197**:1110-1116

[109] Christoffers MJ, Berg ML, Messersmith CG. An isoleucine to leucine mutation in acetyl-CoA carboxylase confers herbicide resistance in wild oat. Genome. 2002;**45**: 1049-1056

[110] Moss SR, Cocker KM, Brown AC, Hall L, Field LM. Characterisation of target-site resistance to ACCase-inhibiting herbicides in the weed *Alopecurus myosuroides* (black-grass). Pest Management Science. 2003;**59**:190-201

[111] Yu Q, Collavo A, Zheng MQ, Owen M, Sattin M, Powles SB. Diversity of Acetyl-Coenzyme a carboxylase mutations in resistant *Lolium* populations: Evaluation using clethodim. Plant Physiology. 2007;**145**:547-558

[112] Délye C, Matéjicek A, Michel S. Cross-resistance patterns to ACCase-inhibiting herbicides conferred by mutant ACCase isoforms in *Alopecurus myosuroides* Huds. (black-grass), re-examined at the recommended herbicide field rate. Pest Management Science. 2008;**64**:1179-1186

[113] Scarabel L, Panozzo S, Varotto S, Sattin M. Allelic variation of the ACCase gene and response to ACCase-inhibiting herbicides in pinoxaden-resistant *Lolium* spp. Pest Management Science. 2011;**67**:932-941

[114] Beckie HJ, Warwick SI, Sauder CA. Basis for herbicide resistance in Canadian populations of wild oat (*Avena fatua*). Weed Science. 2012;**60**(1):10-18

[115] Papapanagiotou AP, Paresidou MI, Kaloumenos NS. Eleftherohorinos IG. ACCase mutations in *Avena sterilis* populations and their impact on plant fitness. Pesticide Biochemistry and Physiology. 2015;**123**:40-48

[116] Saini RK, Malone J, Preston C, Gill G. Target enzyme-based resistance to clethodim in *Lolium rigidum* populations in Australia. Weed Science. 2015;**63**:946-953

[117] Keshtkar E, Mathiassen SK, Moss SR, Kudsk P. Resistance profile of herbicide-resistant *Alopecurus myosuroides* (black-grass) populations in Denmark. Crop Protection. 2015;**69**:83-89

[118] Iwakami S, Hashimoto M, Matsushima K, Watanabe H, Hamamura K, Uchino A. Multiple-herbicide resistance in *Echinochloa crus-galli* var. *formosensis*, an allohexaploid weed species, in dry-seeded rice. Pesticide Biochemistry and Physiology. 2015;**119**:1-8

[119] Guo W, Liu W, Li L, Yuan G, Du L, Wang J. Molecular basis for resistance to Fenoxaprop in Shortawn Foxtail (*Alopecurus aequalis*) from China. Weed Science 2015;63 416-424

[120] Cocker KM, Moss SR, Coleman JOD. Multiple mechanisms of resistance to fenoxaprop-P-ethyl in UK and other European populations of herbicide-resistant *Alopecurus myosuroides* (black-grass). Pesticide Biochemistry and Physiology. 1999;**65**:169-180

[121] Letouzé A, Gasquez J. Inheritance of fenoxaprop-P-ethyl resistance in a blackgrass (*Alopecurus myosuroides* Huds.) population. Theoretical and Applied Genetics. 2001;**103**: 288-296

[122] Petit C, Duhieu B, Boucansaud K, Délye C. Complex genetic control of non-target-site-based resistance to herbicides inhibiting acetyl-coenzyme A carboxylase and acetolactate-synthase in *Alopecurus myosuroides* Huds. Plant Science. 2010;**178**:501-509

[123] Cocker KM, Coleman JOD, Blair AM, Clarke JH, Moss SR. Biochemical mechanisms of cross-resistance to aryloxyphenoxypropionate and cyclohexanedione herbicides in populations of *Avena* spp. Weed Research. 2000;**40**:323-334

[124] Ahmad-Hamdani MS, Yu Q, Han H, Cawthray GR, Wang SF, Powles SB. Herbicide resistance endowed by enhanced rates of herbicide metabolism in wild oat (*Avena* spp.). Weed Science. 2013;**61**:55-62

[125] Maneechote C, Preston C, Powles SB. A diclofop-methyl-resistant Avena sterilis biotype with a herbicide-resistant acetyl-coenzyme A carboxylase and enhanced metabolism of diclofop-methyl. Pesticide Science. 1997;**49**:105-114

[126] Li L, Bi Y, Liu W, Yuan G, Wang J. Molecular basis for resistance to fenoxaprop-p-ethyl in American sloughgrass (*Beckmannia syzigachne* Steud.). Pesticide Biochemistry and Physiology. 2013;**105**:118-121

[127] Tang W, Zhou Z, Zhang Y, Chen J. Resistance of American sloughgrass (*Bechmannia syzigachne*) populations to ACCase-inhibiting herbicides involves three different target site mutations from China. Pesticide Biochemistry and Physiology. 2015;**124**:93-96

[128] Huan ZB, Jin T, Zhang SY, Wang JX. Cloning and sequence analysis of plastid acetyl-CoA carboxylase cDNA from two *Echinochloa crus-galli* biotypes. Journal of Pesticide Science. 2011;**36**(4):461-466

[129] Huan Z, Xu Z, Lv D, Wang J. Determination of ACCase Sensitivity and gene expression in Quizalofop-Ethyl-Resistant and -Susceptible barnyardgrass (*Echinochloa crus-galli*) biotypes. Weed Science. 2013;**61**:537-542

[130] Wright AA, Nandula VK, Grier L, Showmaker KC, Bond JA, Peterson DG, Ray JD, Shaw DR. Characterization of Fenoxaprop-P-Ethyl–Resistant junglerice (*Echinochloa colona*) from Mississippi. Weed Science. 2016;**64**:588-595

[131] McCullough PE, McElroy JS, Yu J, Zhang H, Miller TB, Chen S, Johnston CR, Czarnota MA. ALS-resistant spotted spurge (*Chamaesyce maculata*) confirmed in Georgia. Weed Science. 2016;**64**:216-222

[132] Shergill LS, Malone J, Boutsalis P, Preston C, Gill G. Target-Site point mutations conferring resistance to ACCase-inhibiting herbicides in smooth barley (*Hordeum glaucum*) and hare barley (*Hordeum leporinum*). Weed Science. 2015;**63**:408-415

[133] Matthews N, Powles SB, Preston P. Mechanisms of resistance to acetyl-coenzyme A carboxylase-inhibiting herbicides in a *Hordeum leporinum* population. Pest Management Science. 2000;**56:**441-447

[134] White GM, Moss SR, Karp A. Differences in the molecular basis of resistance to the cyclohexanedione herbicide sethoxydim in *Lolium multiflorum.* Weed Research. 2005;**45**:440-448

[135] Kaundun SS, Hutchings SJ, Dale RP, McIndoe E. Broad resistance to ACCase inhibiting herbicides in a ryegrass population is due only to a cysteine to arginine mutation in the target enzyme. PLoS ONE. 2012;**7**(6):e39759

[136] Cocker KM, Northcroft DS, Coleman JOD, Moss SR. Resistance to ACCase-inhibiting herbicides and isoproturon in UK populations of *Lolium multiflorum* mechanisms of resistance and implications for control. Pest Management Science. 2001;**57**:587-597

[137] Martins BAB, Sánchez-Olguín E, Perez-Jones A, Hulting AG, Mallory-Smith C. Alleles contributing to ACCase-resistance in an Italian ryegrass (*Lolium perenne* ssp. *multiflorum*) population from oregon. Weed Science. 2014;**62**:468-473

[138] Collavo A, Panozzo S, Allegri A, Sattin M. A "Stressed" Alfalfa-Based cropping system leads to the selection of Quizalofop-Resistant Italian ryegrass (*Lolium perenne* ssp. *multiflorum*). Weed Science. 2016;**64**:683-694

[139] Preston C. Inheritance and linkage of metabolism-based herbicide cross-resistance in rigid ryegrass (*Lolium rigidum*). Weed Science. 2003;**51**:4-12

[140] Busi R, Vila-Aiub MM, Powles SB. Genetic control of a cytochrome P450 metabolism based herbicide resistance mechanism in *Lolium rigidum*. Heredity. 2011;**106**:817-824

[141] Yuan G, Liu W, Bi V, Du L, Guo W, Wang J. Molecular basis for resistance to ACCase-inhibiting herbicides in *Pseudosclerochloa kengiana* populations. Pesticide Biochemistry and Physiology. 2015;**119**:9-15

[142] Déley C, Wang T, Darmency H. An isoleucine-leucine substitution in chloroplastic acetyl-CoA carboxylase from green foxtail (*Setaria viridis* L. Beauv.) is responsible for resistance to the cyclohexanedione herbicide sethoxydim. Planta. 2002;**214**:421-427

[143] Marles MAS, Devine MD, Hall CJ. Herbicide resistance in *Setari aviridis* conferred by sell sensitive forms of acetyl coenzyle A carboxylase. Pesticide Biochemistry and Physiology. 1993;**46**:7-14

[144] Scarabel L, Panozzo S, Savoia W, Sattin M. Target-site ACCase-resistant johnsongrass (*Sorghum halepense*) selected in summer dicot crops. Weed Technology. 2014;**28**(2): 307-315

[145] Burke IC, Burton JD, York AC, Cranmer J, Wilcut JW. Mechanism of resistance to clethodim in a johnsongrass (*Sorghum halepense*) biotype. Weed Science. 2006;**54**(3):401-406

[146] Steinrucken HC, Amrhein N. The herbicide glyphosate is a potent inhibitor of 5-enol-pyruvylshikimic acid-3-phosphate synthase. Biochemical and Biophysical Research Communications. 1980;**94**:1207-1212

[147] Baylis AD. Why glyphosate is a global herbicide: Strengths, weaknesses and prospects. Pest Management Science. 2000;**56**:299-308

[148] James C, Krattiger AF. Global Review of the Field Testing and Commercialization of Transgenic Plants, 1986 to 1995: The First Decade of Crop Biotechnology. ISAAA Briefs No. 1. Ithaca, NY: ISAAA; 1996. p. 31

[149] Bradshaw LD, Padgette SR, Kimball SL, Wells BH. Perspectives on glyphosate resistance. Weed Technology. 1997;**11**:189-198

[150] Nandula KV, Ray DJ, Ribeiro ND, Pan Z, Reddy NK. Glyphosate resistance in tall waterhemp (*Amaranthus tuberculatus*) from Mississippi is due to both altered target-site and nontarget-site mechanisms. Weed Science. 2013;**61**:374-383

[151] Chatham AL, Bradley WK, Kruger RG, Martin RJ, Owen DKM, Peterson ED, Mithila J, Tranel JP. A multistate study of the association between glyphosate resistance and EPSPS gene amplification in waterhemp (*Amaranthus tuberculatus*). Weed Science. 2015;**63**:569-577

[152] Mohseni-Moghadam M, Schroeder J, Ashigh J. Mechanism of resistance and inheritance in glyphosate resistant palmer amaranth (*Amaranthus palmeri*) populations from New Mexico, USA. Weed Science. 2013;**61**:517-525

[153] Young BG, Knepp AW, Wax LM, Hart SE. Glyphosate translocation in common lambsquarters (*Chenopodium album*) and velvetleaf (*Abutilon theophrasti*) in response to ammonium sulfate. Weed Science. 2003;**51**:151-156

[154] Krga I, Pavlović D, Anđelković A, Đurović S, Marisavljević D. The resistance of orchard weed populations to glyphosate (In Serbian: Otpornost korovskih populacija u voćnjacima na glifosat). Plant Protection. 2013;**64**(3):125-133

[155] Boerboom MC, Wzese LD. Influence of glyphosate concentration on glyphosate absorption and translocation in Canada Thistle (*Cirsium arvense*). Weed Science. 1988;**36**: 291-295

[156] Bourque J, Chen YS, Heck G, Hubmeier C, Reynolds T, Tran M, Ratliff PG, Sammans D. Investigation into glyphosate-resistant horseweed (*Conyza canadensis*): Resistance mechanism studies. Book of Abstract of the Weed Science Society of America. 2002;**42**:65

[157] Feng PC, Tran M, Chiu T, Sammons RD, Heck GR, CaJacob CA. Investigations into glyphosate resistant horseweed (*Conyza canadensis*): retention, uptake, translocation and metabolism. Weed Science. 2004;**52**:498-505

[158] Koger CH, Reddy KN. Role of absorption and translocation in the mechanism of glyphosate resistance in horseweed (*Conyza canadensis*). Weed Science. 2005;**53**:84-89

[159] Dinelli G, Marotti I, Bonetti A, Minelli M, Catizone P, Barnes J. Physiological and molecular insight on the mechanisms of resistance to glyphosate in *Conyza canadensis* (L.) Cronq biotipes. Pesticide Biochemistry and Physiology. 2006;**86**:30-41

[160] Pavlovic D, Reinhardt FC, Bozic D, Vrbnicanin S. Determination of *Conyza canadensis* levels of sensitivity to glyphosate trmesium sulphsate. International Journal of Agriculture and Biology. 2013;**15**(6):1091-1097

[161] Dinelli G, Marotti I, Bonetti A, Catiyone P, Urbano JM, Barnes J. Physiological and molecular mechanisms of glyphosate resistance in *Conyza bonariensis* (L.) Cronq biotypes from Spain. Weed Research. 2008;**48**:257-265

[162] Yerka MK, De Leon N, Stoltenberg DE. Pollen-mediated gene flow in common lambsquarters (*Chenopodium album*). Weed Science. 2012;**60**(4):600-606

[163] Tehranchian P, Riar DS, Norsworthy JK, Nandula V, McElroy S, Chen S, Scott RC. ALS-resistant smallflower umbrella sedge (*Cyperus difformis*) in arkansas rice: Physiological and molecular basis of resistance. Weed Science. 2015;**63**:561-568

[164] Wang YC. Effect of glyphosate on aromatic amino acid metabolism in purple nutsedge (*Cyperus rotundus*). Weed Technology. 2001;**15**(4):628-635

[165] Baerson SR, Rodriguez DJ, Tran M, Feng Y, Biest NA, Dill GM. Glyphosate-Resistant goosegrass. Identification of a mutation in the target enzyme 5-Enolpyruvylshikimate-3-Phosphate synthase. Plant Physiology. 2002;**129**:1265-1275

[166] Ng CH, Wickneswari R, Salmijan S, Teng YT, Ismail BS. Gene polymorphisms in glyphosate-resistant and susceptible biotypes of *Eleusine indica* from Malaysia. Weed Research. 2003;**43**:108-115

[167] Kaundun SS, Zelaya IA, Dale RP, Lycett AJ, Carter P, Sharples KR, McIndoe E. Importance of the P106S target-site mutation in conferring resistance to glyphosate in a goosegrass (*Eleusine indica*) population from the Philippines. Weed Science. 2008;**56**:637-646

[168] Alarcón-Reverte R, García A, Urzúa J, Fischer AJ. Resistance to glyphosate in junglerice (*Echinochloa colona*) from California. Weed Science. 2013;**61**:48-54

[169] Wakelin AM, Preston CA. Target-site mutation is present in a glyphosate-resistant *Lolium rigidum* population. Weed Research. 2006;**46**:432-440

[170] Kaundun SS, Dale RP, Zelaya IA, Dinelli G, Marotti I, McIndoe E, Cairns A. A novel P106L mutation in EPSPS and an unknown mechanism(s) act additively to confer resistance to glyphosate in a South African *Lolium rigidum* population. Journal of Agricultural and Food Chemistry. 2011;**59**(7):3227-3233

[171] Feng PCC, Pratley JE, Bohn JA. Resistance to glyphosate in *Lolium rigidum*. II Uptake, translocation and metabolism. Weed Science. 1999;47:412-415

[172] Simarmata M, Kaufmann JE, Penner D. Potential basis of glyphosate resistance in California rigid ryegrass (*Lolium rigidum*). Weed Science. 2003;**51**:678-682

[173] Lorraine-Colwill DF, Powles SB, Hawkes TR, Hollinshead PH, Warner SAJ, Preston C. Investigations into the mechanism of glyphosate resistance in *Lolium rigidum*. Pesticide Biochemistry and Physiology. 2003;**74**:62-72

[174] Wakelin AM, Lorraine-Colwill DF, Preston C. Glyphosate resistance in four different populations of *Lolium rigidum* is associated with reduced translocation of glyphosate to meristematic zones. Weed Research. 2004;**44**:453-459

[175] Adu-Yeboah P, Malone MJ, Gill G, Preston C. Reduced glyphosate translocation in two glyphosate-resistant populations of rigid ryegrass (*Lolium rigidum*) from fence lines in South Australia. Weed Science. 2014;**62**:4-10

[176] Baerson SR, Rodriguez DJ, Biest NA, Tran M, You J, Kreuger RW, Dill GM, Pratley JE, Gruys KJ. Investigating the mechanism of glyphosate resistance in rigid ryegrass (*Lolium rigidum*). Weed Science. 2002;**50**:721-730

[177] Pavlović D, Reinhardt C, Elezović I, Vrbničanin S. Identification of glyphosate resistance in *Lolium rigidum* Gaudin. Pesticides and Phytomedicine. 2011;**26**(4):393-399

[178] Perez-Jones A, Park KW, Polge N, Colquhoun J, Mallory-Smith CA. Investigating the mechanisms of glyphosate resistance in *Lolium multiflorum*. Planta. 2007;**226**:395-404

[179] Michitte P, De Prado R, Espinoza N, Ruiz-Santaella JP, Gaurit C. Mechanisms of resistance to glyphosate in a Ryegrass (*Lolium multiflorum*) biotype from Chile. Weed Science. 2007;**55**(5):435-440

[180] Salas AR, Scott CR, Dayan EF, Burgos RN. EPSPS Gene Amplification in glyphosate-resistant Italian ryegrass (*Lolium perenne* ssp. *multiflorum*) populations from Arkansas (United States). Journal of Agricultural and Food Chemistry. 2015;**63**:5885-5893

[181] Cross BR, McCarty BL, Tharayil N, McElroy SJ, Chen S, McCullough EM, Powell AB, Bridges CW. A Pro106 to Ala substitution is associated with resistance to glyphosate in annual bluegrass (*Poa annua*). Weed Science. 2015;**63**:613-622

[182] Riar DS, Norsworthy JK, Johnson DB, Scott RC, Bagavathiannan M. Glyphosate resistance in a johnsongrass (*Sorghum halepense*) biotype from arkansas. Weed Science. 2011;**59**:299-304

[183] Norsworthy JK, Ward SM, Shaw DR, Lewellyn RS, Nichols RL, Webster TM, Bradley KW, Frisvold G, Powles SB, Burgos NR, Witt WW, Barrett M. Reducing the risks of herbicide resistance: Best management practices and recommendations. Weed Science 2012;**12**(Special Issue):31-62

Chapter 3

# Gene Flow from Herbicide-Resistant Crops to Wild Relatives

Sava Vrbničanin, Dragana Božić and
Danijela Pavlović

Additional information is available at the end of the chapter

http://dx.doi.org/10.5772/67645

**Abstract**

Development of herbicide-resistant (HR) crops is way to overcome problems in weed control due to weed resistance to herbicides and absence of new herbicides with a new mode of action for their control. Three types of HR crops were developed: nontransgenic, transgenic, and multiple HR crops. Cultivation of HR crops is associated not only with many benefits (simplification of weed control, more effective and efficient weed control, higher yields, etc.) but also with various risks (development of HR weeds, development of HR volunteer crops, gene flow from HR crops to susceptible relatives, etc.). The greatest risk is gene flow from HR crops to related weed species, wild relatives or conventional crops of the same species. Unwanted gene flow could be prevented or reduced using different barriers such as isolation in space or time, protective vegetation barriers, male sterility, etc. Sunflower hybrids resistant to herbicides (imidazolinones and sulfonilureas) was developed by conventional breeding methods, and their introduction in Serbian fields has enabled a more efficient control of harmful weed species, but the presence of huge populations of weedy sunflower is the main concern associated with their cultivation, because numerous studies have confirmed gene flow from sunflower to its relatives.

**Keywords:** gene flow, herbicide resistant crops, wild relatives

## 1. Introduction

The main aim of plant breeding is creating new varieties and hybrids, which would enable us to overcome different problems of contemporary agriculture and achieve high yields and productivity. Research in the fields of molecular genetics, biochemistry, and physiology is leading to development of plants with additional agronomic properties, such as herbicide resistance, pathogen and pest resistance, salt and dryness tolerance, certain food quality parameters, etc. [1–4]. The predominant resistances used in crops are herbicide resistance, both in nontransgenic and in transgenic crops. Owing to the novel insights into the mechanisms and site of action of herbicides on a molecular level, and the development of new biotechnology methods, breeding of herbicide-resistant (HR) crops has been enabled. Thanks to that it is possible to use herbicides, which are preferable from agronomic, environmental, or genetic viewpoint. This new biotechnology gives many benefits in food production such as higher yield through high efficiency of weed control, less unit cost of food production, better quality through removal of existing volunteers of the some species, the possibility of using low-tillage systems, etc. But, this new biotechnology also has some disadvantages such as development herbicide-resistant weed species due to high selection pressure, potential for development of herbicide-resistant volunteer crops, risks of cross-pollination and gene flow from resistant to susceptible relatives, etc.

The focus of this chapter is review of risks associated with HR crops growing with special attention on gene flow from crops to their wild relatives. We first discuss development of HR crops and technologies of weed control based on resistant crops. Also, we briefly discuss gene flow from HR crops to their wild relatives and barriers, which can prevent it. Finally, we discuss transfer of genes responsible for resistance from sunflower hybrids (present resistant crop in Serbia and in Europe) to wild sunflower forms.

## 2. Herbicide-resistant crops

Discovery of new herbicides, especially with a new mode of action is difficult and expensive. During the last few decades, no one herbicide with novel site of action was found and there are no expectations for its appearance in the near future [5, 6]. One way to overcome this problem was development HR crops, which provide expanding the utility of existing herbicides and improve weed control with them. The study on developing HR crops started soon after the discovery of first herbicide-resistant weeds [7, 8]. These type of crops are designed to tolerate specific broad-spectrum herbicides, which kill the surrounding weeds, but leave the cultivated crop intact. There were two directions in HR crops development, which resulted with two groups of crops: transgenic (genetically modified, GM) and nontransgenic HR crops. The first nontransgenic program for HR crops breeding transferred resistance to herbicide triazines from a *Brassica rapa* to canola [9]. Although several triazine-resistant canola varieties were developed, farmer interest for these varieties was poor due to pleiotropic effects of mutation responsible for resistance, which caused lower yielding and poorer seedling vigor [10]. Also, nontransgenic methods like whole-cell selection, mutagenesis, and plant selection from natural populations have been used for breeding of crops resistant to sulfonylurea, sethoxydim, and imidazolinone herbicides. At the same period (1980s), tools for producing transgenic crops were becoming available and many companies start to work on their development. Bromoxynil-resistant cotton was one of the first transgenic HR crops available to farmers in 1995 [11], followed by glyphosate-resistant maize, canola, cotton, soybean, and other crops known as "Roundup Ready" crops. After period of effective, simple, and inexpensive weed

http://dx.doi.org/10.5772/67645

management with cultivation glyphosate-resistant crops, glyphosate-resistant weeds becoming a problem in weed control, which increase the use of crops resistant to glufosinate [12], followed by initiation of new approach in HR crops development, which was based on building of multiple resistance in crop plants.

Significant number of crop plants resistant to different ALS (acetolactate synthase; also known as AHAS—acetohydroxyacid synthase) inhibiting herbicides were developed using conventional breeding methods (**Table 1**). These groups of herbicides have very good characteristics for utilization in weed control in HR crops, which include low use rates, broad spectrum weed control, low mammalian toxicity and environmental compatibility. Immediately after discovery of this group of herbicides, ALS resistant tobacco and maize lines were developed using tissue culture selection [13, 14], while ALS-resistant soybean developed using mutagenesis [15].

| | **Herbicide** | **Crop** | **First market** |
|---|---|---|---|
| Non-transgenic | Photosystem II inhibitors | Soybean | ~1991 |
| | | Canola | 1984 |
| | Imidazolinones | Maize | 1992 |
| | | Canola | 1995 |
| | | Wheat | 2001 |
| | | Rice | 2001 |
| | | Sunflower | 2003 |
| | Sulfonylureas | Soybean | 1994 |
| | | Sunflower | 2006 |
| | | Sorghum | ~2013 |
| | ACCase inhibitor sethoxydim | Maize | 1996 |
| Transgenic | Glyphosate | Soybean | 1996 |
| | | Canola | 1996 |
| | | Cotton | 1997 |
| | | Maize | 1998 |
| | | Alfalfa | 2006 |
| | | Sugarbeet | 2007 |
| | Glufosinate | Canola | 1995 |
| | | Maize | 1996 |
| | | Cotton | 2005 |
| | | Rice | 2006 |
| | | Soybean | 2009 |

**Table 1.** Some commercialized HR crops, modified from reference [20].

After that, three technologies of weed control, which include crop resistance to this group of herbicides, were developed. The Clearfield® and the Clearfield Plus® system have been developed with the aim to grow crops resistant to IMI herbicides [16], while ExpresSun® system has been developed with the aim to grow sunflower hybrids resistant to tribenuron-methyl [17]. As there is no "alien" genes introduced into these crops, this group of HR crops is not considered as transgenic and has been accepted in countries where the cultivation of GM crops is prohibited [16], like many European countries, as well as in Serbia.

Transgenic (GM) crops developed based on the use of different transgenes, mainly responsible for resistance to glyphosate, which introduced into many crop species (**Table 1**). These crops became popular thanks to simplification of weed control and reduction of production costs, making the crop more profitable. Between more than a hundred GM products, which have been authorized for commercialization only 13 are crops [18]. The main GM crops are maize, soybean, cotton, and rapeseed, which grow on more than 90 million ha distributed in 14 countries in which these crops have been authorized [19]. These crops are grown in America, Australia, China, South Africa, but distribution is the highest in the USA, where it covers more than 49.8 million ha [19]. In Europe, GM crops (maize, rapeseed, endive, soybean, and flowers) adopted for the production and/or consumption only in few countries, between which Spain is major producer, growing GM maize on more than 100,000 ha [19].

New approach in development of HR crops is technology, which combines glyphosate resistance with resistance to other herbicides resulting in multiple HR crops (**Table 2**). This technology developed with the aim to overcome increasing development of multiple HR weeds and based on engineering crops that are able to express multiple HR traits and tolerate multiple herbicides. This new concept using stacked (contains more than one transgene) genes as a tool for postoccurrence and future resistance management is the equivalent to using a single herbicide in case when weed is already resistant to one member of a dual stack [21]. Appropriate transgene stacks should delay resistance longer than approach, which use each component separately and sequentially because each weed resistant to either herbicide will be killed by the other herbicide in the stack. However, that stacking multiple HR into crops may or may not delay the evolution of herbicide resistance because effectiveness of the transgene stacks depends on the management decisions

| **Herbicide types** | **Crops** |
|---|---|
| Glyphosate and glufosinate | Soybean, maize, cotton |
| Glyphosate and ALS inhibitors | Soybean, maize |
| Glyphosate, glufosinate and 2,4-D analogs | Soybean, cotton |
| Glyphosate, glufosinate and dicamba | Soybean, cotton |
| Glyphosate, glufosinate and HPPD inhibitors | Soybean, cotton |
| Glyphosate, glufosinate, 2,4-D and ACCase inhibitors | Maize |

**Table 2.** Multiple HR crops under development [20].

and adoption of the accompanying stewardship programs [21]. Namely, it depends on the effectiveness of each included herbicide in control of each target weed species. Some soybean multiple resistant cultivars have recently been approved for commercial use, such as cultivars resistant to glyphosate, glufosinate, and 2,4-D, as well as resistant to glyphosate and dicamba [21]. Except that it is possible to develop stacks of transgenes for different traits. For example, maize containing transgenes for resistance to insects and to herbicides is commercialized [22].

## 3. Benefits and risks associated with growing of herbicide-resistance crops

Cultivation of HR crops is associated not only with numerous benefits but also with various risk factors. The most important benefit is simplification of weed control using herbicides (including nonselective herbicides in many HR crops), in which some crops are able to control weeds that other herbicide that cannot control without concern for crop injury. Also, HR crops are good solution for control of parasitic weed species, in which control is more complex due to their attachment for host (mainly crop) plants [23]. Thanks to flexibility to the time of herbicide application, possible combination with other herbicides and integration with nonchemical methods, this weed control approach made weed management more effective and efficient, which results in higher and more profitable yields. For example, the average increase of yield of glyphosate-resistant soybean in developed countries was 7%, while in developing countries, it was 21% [24]. The higher yield with better quality of seed is not a direct result of HR crop traits *per se*, but it is the result of improved weed control, which is mainly more effective than the conventional weed management systems [20]. These approach to weed control also became popular thanks to the absence of new herbicides with novel sites of action during the last few decades and no prediction for its appearance in the near future [5, 6]. Also, weed resistance to herbicides becomes widely spread and still growing problem, which is difficult to manage. In conventional weed management systems control of weeds which are closely related to the crop is difficult or impossible. ALS-inhibiting resistant sunflower would allow to use this group of herbicides to control *Ambrosia trifida, Ambrosia artemisiifolia, Cirsium arvense, Xanthium strumarium*, weedy *Helianthus annuus*, and other weeds belonging Asteraceae family, without injuring the crop [25]. Although there are controversial views about HR crops impacts on environment, it is evident that this weed control system is more beneficial to the environment than conventional systems. Namely, herbicides used in HR crops are usually more environmentally friendly than herbicides used in conventional crops. ALS-inhibiting herbicides, which are used in many nontransgenic HR crops, are very effective with relatively low use rates and low mammalian toxicity [26]. Also, it is clearly shown that glyphosate-resistant crops are beneficial to the environment by reducing fuel use and soil erosion and residues of herbicides in ground waters [20] with the help of reduced tillage. Also, glyphosate-resistant crop cultivation has decreased herbicide use by 17 million kg per yr in the USA [27]. Except described, there are additional benefits growing HR crops to which farmers also give great importance. For example, in case of glyphosate-resistant maize, soybean and cotton, growers highlighted as very important (**Figure 1**) consistency and protection from yield loss, application frequency and flexibility, a clean field, cost, crop safety, family and public health, water quality, etc. [28].

The cultivation of HR crops, whether they have been developed through genetic engineering or classical breeding methods, is fraught with risks, i.e., potential serious economic and ecological consequences. Unlike the HR crops, which have been obtained through conventional breeding methods, the cultivation of GM crops has been a cause of a number of debates, pertaining to the health safety of these products and the risks they present to the environment. The questions, which cause the greatest concern, are those which relate to: (1) direct and indirect toxic effects of products containing transgenes for nonspecific organisms; (2) the impact of modified genes and GM plants on biodiversity, ecosystems, and soil microorganisms; and (3) gene transfer from GM crops to their wild relatives and ecological consequences of this phenomenon [29]. Contrary to this, in the case of HR crops developed by conventional breeding methods, the greatest risk is transfer of genes responsible for resistance from those crops to related weed species, wild relatives, or conventional crops of the same species. Namely, described gene flow creating the hybrids between HR crops and weeds, the so-called "super weeds", resistant to herbicides. Their eradication subsequently becomes one of the major problems in agriculture. Also, gene flow can change the fitness of recipient biotype/species, whereby increase of fitness resulting in greater weediness, while its decreases lead to extinction [30]. In addition, genes responsible for resistance can flow from HR crops to conventional varieties, which could be the source for resistant genes flow to wild or weedy relatives [31]. Gene flow from transgenic to nontransgenic crops of the same species has been a major controversy, the cause of law suits, and a factor influencing commercialization of some transgenic crops. Some authors [32] highlighted that the risks associated with transgenic crops cultivation may be more pronounced in the centers of origin of crops than in the other territories because of the presence of wild progenitors and other wild relatives in centers of origin.

The occurrences of volunteer populations of HR crops can also be leading to high risk. Namely, seed dissipation during harvest lead to the appearance of volunteer plants the next season

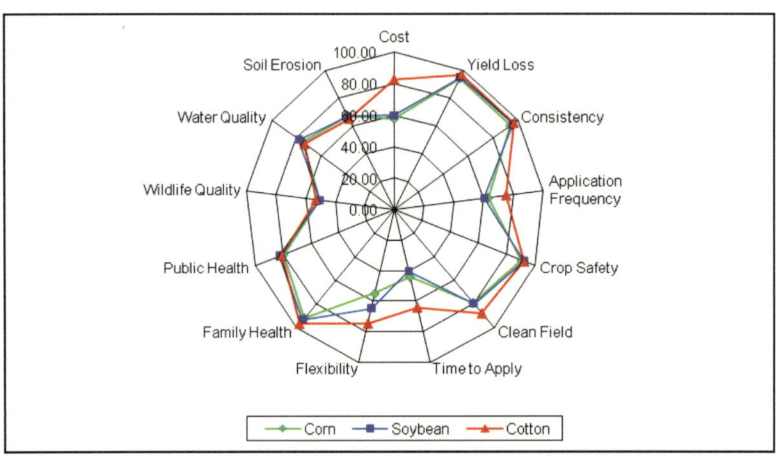


**Figure 1.** The important benefits growing of HR crops from the farmer point of view, made based on survey between 1176 glyphosate-resistant maize, soybean, and cotton growers [28].

http://dx.doi.org/10.5772/67645

generally in crop production of some crops. Negative consequences of volunteer plants are yield and quality reduction of the crop which they have invaded, contamination of harvested seeds, and maintenance of harmful insects and diseases. In the case of HR crops that volunteer plants basically represent resistant weed populations, which can be a source of pollen, which can contaminate the nonresistant crops or pass the resistance traits onto the related weed species. The control of volunteer plants, which has originated from HR crops, is impossible in the following cultures in which these herbicides are applied as a weed control measure. Therefore, volunteer plants of glyphosate-resistant cotton could be a problem in glyphosate-resistant soybean as subsequent crop [33] or volunteer glyphosate-resistant canola and wheat could be problem in weed control in conservation tillage system [34]. Also, seeds from volunteer plants of GM crop can contaminate harvest of conventional subsequent crop [35].

Intensive and repeated use of the same herbicides with the same mode of action in HR crops mainly increase selection pressure on weeds, which would most likely lead to an increase in the selection of HR weed populations. Today, at least 36 weed species have evolved resistance to glyphosate, EPSPS inhibitor (the main herbicide in transgenic HR crops), and at least 159 to ALS-inhibiting herbicides (the main group of herbicides in nontransgenic HR crops) [36]. In addition to these concerns, other negative effects are also possible: herbicide drift can damage conventional crops of the same species, the genes responsible for resistance can be transferred onto conventional crops, characteristics of nontarget plants can be modified, biodiversity may be damaged, and the environment and soil properties can be changed due to the changes in the crop production technologies.

Due to the dangers of the mentioned potential risks, the research into these issues, with the aim of developing suitable prevention strategies, as well as solutions to these problems, should they arise, has been intensified. Consequently, plenty have dealt with the issue of the gene transfer from HR crops to their relatives (wild/weedy forms or conventional crops) [37–42], the study of gene stability in recipients [43, 44], the study of crop-weed hybrid's fitness [41, 45–47] and the competition between crop-weed hybrids and sensitive weed plants of the same species [45, 46, 48].

## 4. Gene flow from herbicide-resistant crops to wild or weedy relatives

Hybridization and introgression are normal processes, which have continuously occurred between crops and wild or weedy relatives [49, 50], as well as between relative populations of weedy and/or wild species [51, 52]. Even though the hybridization of crops and weeds has an important role in the evolution of many weed species [53], it can also result in the extinction of certain species related to the crops or the rise of new weed forms, which are more aggressive and better adapted to artificial habitats [30]. There are three types of gene flow: *vertical* (between sexually compatible individuals), *horizontal* (between distant related species), and *diagonal* (between related but incompletely incompatible species) [54], but introgression of genes from cultivated to wild or weedy forms of the same species is possible through *vertical* and *diagonal* gene flow.

The ecological consequences of gene transfer from crops to their wild relatives are determined by the quantity of genes, which are being transferred into the populations of wild plants and weeds and the phenotypic characteristics controlled by these genes. Some of the characteristics are insignificant for the fitness of wild relatives, while others (herbicide resistance, disease resistance, and tolerance to the environmental stress factors) mostly improve it. For example, the first generation crop-wild hybrids produced through hybridization between cultivated and wild radish populations [53, 55, 56] was relatively fecund, produced large quantities of seeds and rapidly evolved increased pollen fertility. Contrary to this, if the introduced genes weaken the fitness of their wild relatives, their invisibility will also decrease. This process can be accelerated by introgression and the introduction of new genes from neighboring crops, which ultimately leads to the extinction of the initial populations of wild relatives [57]. Except ecological consequences, gene flow from crops to weedy relatives is associated with many problems in crop production. Namely, the development of HR crops has given rise to the situation where the hybridization is often seen as a problem, particularly when it relates to the hybridization between GM crops and related species. Also, it is important to bear in mind that in some countries coexist different cropping systems, which cultivate conventional, organic, and GM crops. In that situation, there is risk for gene flow between GM and non-GM cultivars through cross-fertilization due to pollen flow between neighboring fields. Progeny of HR crops and weedy/wild relatives or volunteers will be resistant weeds, in which control is difficult.

Genes responsible for crop's herbicide resistance can be spread in the environment as a result of three mechanisms, including gene transfer across a pollen (as a result of allogamy), seeds (as a result of their dispersal) and for perennial species by the vegetative propagules. Potential for pollen-mediated gene flow is higher for both wind and insect pollinated out-crossing crops than for self-pollinated crops [58]. Although gene flow across a pollen is more studied, gene flow by seeds during commerce may be very important for the long-distance dispersal of genes responsible for resistance to herbicides [59]. The both ways of gene flow from HR crops including both GM and conventionally bred HR crops have been confirmed in many cases [37–40, 60, 61].

The transfer of genes from HR crops to their relatives is dependent on multiple factors (**Figure 2**), such as the coexistence and proximity of the crop and its close relatives, their biology and phenology, type of vector, development of F1 generation, which is fertile and capable of survival, the production of fertile subsequent generations, the potential for gene transmission, chromosome recombination and movement of genes of one species into the genome of another, due to introgressive hybridization and gene persistence in volunteer crop populations [58, 62]. Also, in study about gene flow from glufosinate-resistant rice to improved rice cultivars and weedy rice in China, the conclusion was that gene flow depends on the height of pollen recipient plants [63]. They found that the gene flow was lesser if recipients were taller than in situation when they were shorter.

Cross-pollination between HR crops and sexually compatible wild or conventional cultivated crops of the same species is the major pathway for gene escape. Therefore, transfer of genes responsible for HR between sexually compatible individuals is most often done through pollen, whether within the same population or between different populations [38, 64]. This occurrence is dependent on different factors of which autoincompatibility that enhances allogamy in wild forms, environmental conditions (wind speed and direction, temperature, light intensity, and humidity) as well as the type (wind and/or insect) of pollination vector [37, 38, 65, 66]. In addition to this, the crucial role in gene transfer through pollen lies in the coincidence of the flowering period between the HR crop and its wild relatives. Although experimental data suggest

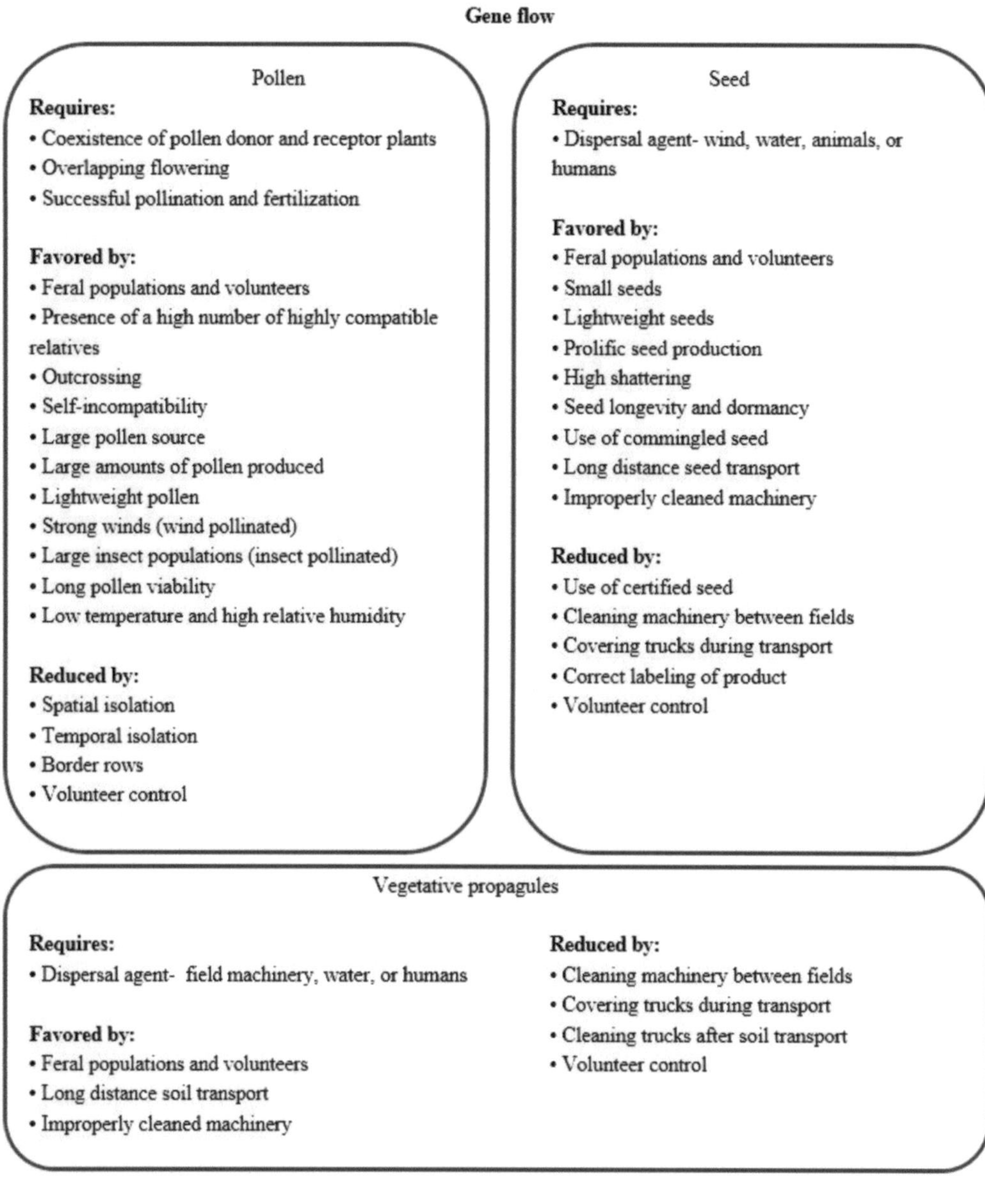


**Figure 2.** Comparison of the requirements and factors affecting gene flow via pollen, seed, and vegetative propagules, modified figure from reference [58].

that the flowering period of wild populations is generally longer than the flowering period of crops, which makes the overlap highly likely [67], in some cases, gene flow between HR crops and relatives was disabled due to flowering period not overlapping or time of overlapping was short. For example, hybridization between imazamox-resistant and weedy sunflower was not confirmed in experiments in Serbia when period of flowering overlapping was short [42]. Also, it was confirmed that the gene transfer from the cultivated onto the wild sunflower in

Argentina depended on the overlap between the flowering period and the presence of common pollinators [68, 69]. Pollen dispersal from HR crops onto their wild relatives is also dependent on their mutual distance, the size of populations from which the pollen originates and where it is delivered, plant density, number of flowers per plant, and the location of wild relatives in relation to the crop [70].

Although numerous studies have confirmed the transfer of genes relevant for HR to their wild relatives, hybridization level mainly was low. Some authors [60] studied the transfer of genes responsible for imazethapyr-resistance, from the rice cultivars to the weedy rice species in 22 field sites. They confirmed that even though gene transfer occurs, in the majority of sites (18) less than 1% of hybrid progeny was present, while in the remaining four sites that percentage was somewhat higher (up to 3%). Also, low levels of hybridization (1–2%) were confirmed between rice and its wild congener *Oryza rufipogon* [71]. Similarly, Ref. [72] confirmed a low level of hybridization between HR rapeseed and related weed species *Raphanus raphanistrum*. Their research has shown that the proportion of crop-weed hybrids in the F1 generation was at the level of $10^{-7}$ to $3 \times 10^{-5}$, depending on the geographic position of the weed species in the experimental plot. Gene flow from glyphosate-resistant canola to *B. rapa* in commercial fields was confirmed, but the genes were apparently not fully introgressed [73]. Contrary to that, in study of transgene escaping from canola to *B. rapa,* the gene frequency in the first backcross generation was 50%. But, in the fourth backcross generation, it was 0.1% in conditions without herbicide application, while in conditions with glyphosate application, gene frequency was about 5.5% within six successive backcross generations [44].

Despite the fact that the gene transfer from crops to their wild relatives is widely studied, there are no detailed data available on what happens with these genes, which have been introduced into wild populations after a longer period of time. Namely, the majority of this research concludes with the first generation of hybrids. However, genes originating from the cultivated sunflower can persist in wild populations over the five-year period, following the hybridization [43]. Some authors [74] have also studied the effects of a 40-year long gene transfer from the cultivated to the wild sunflower populations.

Importance of crop-weed hybrids produced as result of gene flow from HR crops to wild or weedy relatives for future crop production can be different depending on traits introduced into progeny. Therefore, assessment of gene flow occurrence requires not only estimating the degree of gene flow, but also evaluating the relative fitness of hybrids. It long dominated the view that crop-wild hybrids have a lower fitness than their wild parent [75, 76]. But, many studies confirmed that some hybrids display increased [47], while the other display reduced [77] fitness in comparison with their parents. Displayed fitness depends not only on the crop traits introduced to wild relatives, but also on environmental conditions. Namely, fitness of hybrids between crop and wild sunflower increases in stressful conditions common to conventional agroecosystem like competition and herbicide application [77].

The role of seeds in the transfer of HR genes from crops to their wild relatives is evident in their spread into new areas where volunteer populations are formed. After that HR genes can be transferred from these volunteer populations to their wild relatives through the pollen. Also, hybrids resulting from spontaneous crosses of HR crops and their wild relatives through seeds can be carried into new areas, where they subsequently present a source of pollen, which carries the resistance genes. Unlike pollen, the seeds usually remain in the close proximity of the plants from which they have originated. But, as seeds are more persistent than pollen, movement of seeds is possible to further distances by human activities then pollen movement [59]. In general, seed dispersal of HR crops or progeny created through their spontaneous crossing with wild relatives, depends on the biological properties of the crop, the ecological conditions, crop production technology and the agrotechnical measures applied on these fields, following with harvest. Nevertheless, it is possible to monitor the dispersal of these seeds in space and time. Some authors [78] have confirmed the gene transfer of sugar beet to their wild relatives through the seeds whose dispersal resulted from soil transport. Namely, although spontaneous spatial dispersal is often considered as irrelevant since the seeds of a majority of crop cultures have lost this ability, seed dispersal is also possible as a result of spillage during the harvest and their transport and storage operations, which enables the spread to great distances. The dispersal of seeds containing the genes responsible for resistance over time depends on the dormancy characteristics and the seed's longevity in the soil, as well as the ecological requirements for its germination. Also, it should be considered that, in addition to pollen and seeds, soil seedbank has an important role in the plant dispersal [79]. Namely, when considering different life forms of sugar beet (cultivated, wild, and weedy), it is well known that they form long-term seedbanks [80], which, over a longer period of time, can provide the plants which are then a source of HR genes.

Gene flow by the vegetative propagules (stolons, rhizomes, roots, crowns, and bulbs) is possible on short distance via natural means or on equipment moved between fields, while long-distance movement could be possible only with human activities or through the waterways [58]. As HR crops are mainly annual species, gene flow via vegetative propagules can be interesting only for perennials like glyphosate-resistant alfalfa (commercially available) and creeping bentgrass (*Agrostis stolonifera*, under consideration) [58].

*Gene flow risk assessment* is a procedure, which helps determine whether the transfer will occur, and if it will, in which degree, with a goal to reduce such a risk to the minimal possible level. Furthermore, such estimates are also significant due to the possibility that the transfer of genes responsible for HR will lead to an increase in the survival and adaptability of the introduced weed species. Also, it is considered that certain plants can attain the traits of invasive species as a result of introduced genes, making the assessments of long-term consequences of gene transfer from crops to their wild relatives a necessity. There is no same potential for gene flow for all HR crops. For example gene flow from maize is theoretically possible to teosinte, but these species only exist in Mexico and Central America and not yet been reported for contamination with transgenes [81]. On the other hand, there is good potential for introgression from sunflower [74], sugarbeet [82] and rice [83] to wild relatives. Namely, the dangers of the transfer of genes responsible for HR and the ecological consequences of this must be evaluated individually for each specific case (herbicide, plant, wild relatives, etc.), regardless of the fact whether the crop resistance has been achieved through conventional breeding methods or genetic engineering. Crucial steps in the rational assessment of ecological consequences of this phenomenon include the understanding of the following: (1) gene transfer from crops to wild relatives; (2) gene expression and inheritance in hybrids, which have resulted from the gene transfer from crops to wild relatives; (3) changes in fitness in wild relatives caused by the introduction of genes; and (4) the dynamics of the gene transfer from crops to wild populations. The invasibility of crop-weed hybrids, which have originated as a result of gene transfer from HR crops to wild relatives, is dependent on all of these aspects.

Pollen flow from crop to the relative seems as relatively simply process, but gene introgression is complex, occurring in several steps which mean several hybrid generations, which can exchange genes among themselves and coexist many years simultaneously (**Figure 3**). The likelihood of gene transfer from crops to their wild relatives depends on the genetic characteristics of crops and their wild relatives, as well as the homology of their genomes [62]. In the cases where the degree of the homology between the crops and their wild relatives is higher, as in the case of *Beta vulgaris* × *B. maritima* or *Raphanus sativus* × *R. raphanistrum*, the likelihood that the introduction of transferred genes will occur is higher. Additionally, the introduction of genes is dependent on the part of the genome, in which the gene is positioned. Some authors [30] found a possibility of gene introduction from 13 most important crops into wild relatives and determined that 12 of the studied crops can hybridize with their wild relatives. Of the 12 listed crops, cases of introduction have been confirmed for 7, while in the remaining five there is a possibility that the introduction will occur. Also, based on the potential danger of transgenic introgression into their wild relatives, some authors [76] have grouped GM crops based on their risk levels into four categories: high, middle, low, and very low (**Table 3**). A similar categorization pertaining to the risk assessment was also applied by other authors [84, 85].

In order to *prevent or reduce* the unwanted transfer of pollen from HR crops onto their relatives, different barriers can be used, although there is no absolute guarantee that the gene transfer can be prevented in this manner. The most often used barriers are isolation in space or time, protective vegetation barriers made up of one or more different species, male sterility as a genetic mechanism for the prevention of gene transfer, etc.

*Spatial (distance) isolation* means increasing the distance between fields sown by HR crops and populations of its relatives. Also, spatial isolation is applied as preventive measure in production of GM and non-GM crops in coexistence with the aim to avoid contamination products of non-GM crops. It has been known that by increasing the distance between crops

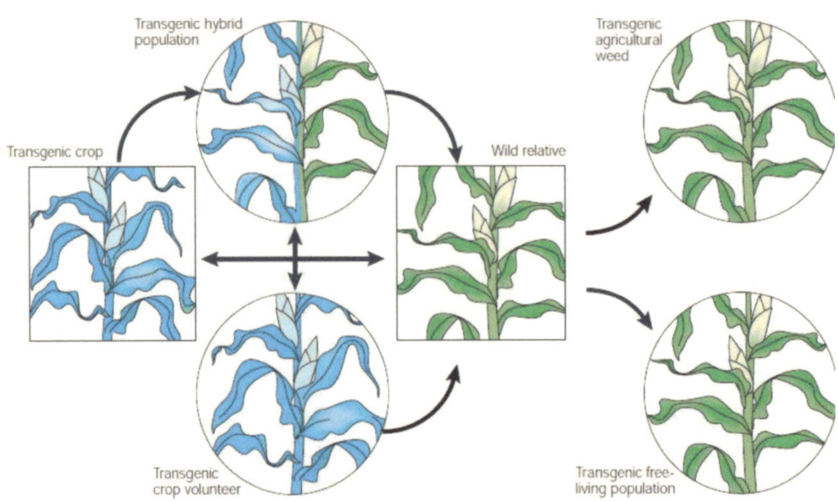


**Figure 3.** Gene flow and its potential resources, modified from reference [76].

| Crop | Risk level | Wild relatives for which the introgression of gene has been confirmed |
|---|---|---|
| Johnson grass | High | *Sorghum halepense, S. almum, S. propinquum* |
| Oilseed rape | Medium | *Brassica rapa, B. juncea, B. oleraceae, B. campestris*<br>*Sinapis arvensis*<br>*Raphanus raphanistrum* |
| Sugar-beet | Medium | *Beta vulgaris* ssp. *vulgaris**<br>*Beta vulgaris* ssp. *maritima* |
| Wheat | Medium | *Triticum turgidum*<br>*Aegilops* sp. (*Aegilops cylindrica*) |
| Sunflower | Medium | *Helianthus* sp. (*H. annuus** and *H. petiolaris*) |
| Alfalfa | Medium | *Medicago sativa** |
| Rice | Low | *Oryza rufipogon* |
| Maize | Low | *Zea mexicana* |
| Potato | Very low | |
| Soybean | Very low | |
| Barley | Very low | |
| Common Bean | Very low | |

*Weedy crop forms.

**Table 3.** The risk level of the introgression of genes from crops to their wild relatives ([86] made based on data reviewed by [76]).

and its relatives, the dispersal of pollen is reduced, i.e., the level of hybridization is reduced. Thus, the frequency of pollen originating from a transgene oilseed rape decrease from 1.5 to 0.00033%, as distance increase from 1 to 47 m [87]. Also, frequency of crop-wild relative hybrid decreased from 0.156 to 0.0038% with increase in distance from pollen source between 200 and 400 m [88]. Distances between pollen source and gene occurrence can be very valuable in the planning of spatial isolation of HR varieties, in order to prevent the gene flow to their relatives. This distance depends on many factors such as the presence of local barriers, the local climate, and the topography of the area. In the case of sunflower, the isolation distance should be greater than 1000 m [89]. Also, maize pollen can be detected at distances greater than 800 m from the pollen source [90]. But, pollen of maize has short flight range [91], after which it settles to the ground rapidly [92] due to relatively heavy and large grains. Due to that cross-fertilization mainly occurs within 50 m of the pollen source [93]. Therefore, measure for keeping seed purity of non-GM maize, which coexists with GM maize, suggests isolation distance between 10 and 50 m to achieve EU admissible threshold of 0.9% in the harvest [37, 93, 94].

*Temporal isolation* is a measure, which should prevent overlapping flowering times of crop and wild relatives with the aim to avoid gene flow. About 5 days lag in flowering of imazamox-resistant in comparison with tribenuron-methyl resistant sunflower resulted in lack of gene flow to weedy sunflower probably due to the short period of overlapping flowering time between the resistant hybrid and the weedy sunflower [42]. Temporal isolation is very suitable to prevent non-GM crop contamination with GM when grow in coexistence. Study of maize pollen mediated gene flow in Italy and showed that if time of flowering differs from 4 to 5 days the cross-pollination is reduced by 25%, while difference of 6 days provides 50% reduction [95]. Also, temporal separation in sowing days improves the coexistence of maize [96]. Contrary to that, temporal isolation based on selection of hybrid varieties in which flowering noncoincide achieves the same results although sowing date was the same [97]. Temporal separation and isolation distance together can be a good solution to minimize unwanted gene flow.

*Protective vegetation barriers*, made up of one or more different species, can prevent the gene flow by physically stopping pollen in the case of wind pollination. The sowing of conventional crops of the same species, also known as pollen traps, in the vicinity of HR hybrids is an efficient measure, as their role is to attract pollinating insects in order to leave pollen on these pollen traps. This type of barrier can be much more effective than isolation distance. Namely, the sowing of pollen traps between GM and conventional crops is the most efficient measure for the prevention of gene flow [98]. Also, gene flow through pollen from the HR oilseed rape decreases rapidly with the increase of pollen source distance, with the added necessity of a protective vegetation barrier or pollen traps [99]. Some authors [95] studied pollen-mediated gene flow between GM and non-GM maize and concluded that effect of two maize rows surrounding the recipient field in reduction of cross-fertilization is the same like effect of 12 maize rows surrounding the pollen donor.

The use of *biological barriers* achieves the best results in the prevention of gene flow, and so far the barriers based on cytoplasmatic male sterility, maternal inheritance, and seed sterility have mostly been used. Cytoplasmic male sterility is based on the inability of plants to produce viable pollen. This type of barrier is suitable option to reduce gene flow in sunflower and maize [43, 66, 100]. Maternal inheritance is successfully used in the prevention of gene flow across the pollen, in the case of several species, including tobacco and tomato [101, 102]. The control of embryo and seed fertility is known as GURT (Gene Use Restriction Technology), i.e., terminator technology, which is considered to be a better control measure, in comparison with sterile pollen production. However, this strategy is seen as the most controversial control measure for limiting genes flow. Additionally, strategies, which include apomixis (vegetative reproduction and asexual seed formation), cleistogamy (self-fertilization without the opening of flowers), genome incompatibility, chemical induction/deletion, etc., are also used in limiting the gene flow [103]. None of these strategies can be applied in all crops, therefore using combinations of different approaches for the prevention of unwanted gene flow is recommended.

All mentioned measures for prevention and reduction of gene flow are important separately, but their integration and combination with stewardship production system could be the best solution.

## 5. Gene flow from herbicide-resistance sunflower to wild or weedy sunflower

Options for chemical control of broadleaf weed species, especially weeds belonging to Asteraceae family, without injuring the crop are quite limited in sunflower compared to most other row crops [104]. Due to that, sunflower hybrids resistant to ALS-inhibiting herbicides, including imidazolinone (IMI) and sulfonylurea (SU), was developed by conventional breeding methods, with the aim to improve weed control. The Clearfield_system [16] and the Clearfield-Plus_system [105] have been developed with the aim to grow sunflower hybrids resistant to IMI herbicides. For development of those hybrids were used for subsequent crossings between cultivated sunflower and wild resistant sunflower [106] or seed mutagenesis [105]. Also, ExpresSun system has been developed as result of mutagenesis breeding [107] with the aim to grow sunflower hybrids resistant to tribenuron-methyl [17].

The breeding of sunflower hybrids resistant to herbicides belonging to IMI and SU groups in Serbia was started in 2000, and since 2003, this technology has been applied in the production. As a donor of imazamox-resistance gene, the wild sunflower originating from the USA was used, in which the resistance to herbicides of the imidazolinone group was developed following a seven-year consecutive application of imazethapyr [106]. The produced hybrid has shown a high level of resistance toward imazethapyr [108] and imazamox [109], not only regarding different vegetative parameters, but also considering the activity of ALS enzymes *in vivo*, and *in vitro*. Source populations SURES-1 and SURES-2 were used as a source of genes responsible for the resistance to tribenuron-methyl [1, 110], producing also a hybrid with a highly distinguished resistance for this herbicide [109, 111]. The introduction of such crops in the production in Serbian fields has enabled a more efficient control of economic harmful weed species, such as *Sorghum halepense*, *A. trifida*, *A. artemisiifolia*, *C. arvense*, *X. strumarium* and weedy forms of *Helianthus annuus*, their cultivation is also linked with a very high risk of herbicide-resistance gene flow, from these hybrids onto the weedy form of *H. annuus*. Although the presence of four species from the genus *Helianthus* (*H. annuus*, *Helianthus tuberosus*, *Helianthus decapetalus*, *Helianthus scaberimus*) has been confirmed for Serbia, in both crop fields and nonarable lands [112], weedy populations of *H. annuus* occupy the biggest areas, which according to some estimates reach up to 1000 ha in Southern Srem and around 7–8000 ha in Southern Banat [113]. The origin of these populations is not known, but it is possible to determine. For example, origin of French and Spanish weedy populations was determined based on molecular analysis, which has shown that these populations originated from the unintentional introduction of crop-wild hybrids through contaminated seed lots [114]. Difficult eradication of weedy populations due to a high population variability [113, 115, 116] and pronounced invasibility caused by strong vegetative and generative potential [117, 118] presents an additional problem. Besides a reduced sensitivity of this species to nicosulfuron, which is often used as a weed control measure in maize fields where weedy sunflower is present in high densities, has also been detected [119, 120]. Therefore, even though the research into the transfer of HR genes from HR sunflower hybrids to weedy sunflower is in initial stages in Serbia [42, 121], there is high potential for its risk.

The main concern associated with cultivation of HR sunflower is potential gene flow from crop to weedy or wild relatives. Although wild sunflower populations are self-incompatible [122], new crop sunflower varieties are about 65% autogamous [123] and weedy population as a result of their hybridization are self-incompatible. Therefore, there is great potential for pollen-mediated gene flow. For example, seed-mediated gene flow from cultivated sunflowers to wild sunflowers may be common [124]. Also, it has been known that there are inter- and intraspecific hybridization between *H. annuus* and its close relatives including its related species [41, 64, 74, 125] or its volunteer plants [65, 69]. In case of gene flow between cultivated sunflower and *Helianthus petiolaris*, the proportion of crop-weed hybrids in the F1 generation varies between 0.3 and 0.5%, depending on flowering period and the presence of common pollinators [68], while in case of gene flow between cultivated and wild sunflower, it was reached as high as 33% [69].

Gene flow from sunflower crops onto their wild relatives mediated by pollen is dependent on different factors. The overlap of flowering periods of cultivated sunflower and its wild relatives, the pollinators which they share, self-incompatibility of the wild species, diploidy, and high levels of cross-fertilization are all factors which contribute to the spontaneous hybridization [66]. However, the hybridization between the sunflower and its relatives can be absent due to the mismatch of the flowering periods, incompatibility, physical distance, differences in the genetic structure between the species and interspecific competition of pollen [89, 125]. Many studies [42, 70, 121] confirmed that the pollen transfer from the resistant crops to their relatives primarily depends on their distance to the pollen source and the plot size. Consequently, some authors [64] have confirmed, when studying gene flow from sunflower imidazolinones-resistant hybrids to their wild relatives, that the HR gene was transported to a distance greater than 30 m from the pollen source, while the percentage of the surviving offspring of wild relatives was reduced with the increase in the distance from the HR hybrid. Also, it has been confirmed that the gene flow from the crop sunflower to its wild form is reduced with an increase in their mutual distance, with it being 27% at a 3 m distance. However, gene flow has also been confirmed at a distance of over 1000 m from the pollen sources [89]. Additionally, it was determined that 42% of the wild offspring sunflower at a 3 m distance from the crop sunflower represented its hybrids, while at a distance of 200 m, this percentage was 10%, and 4% at a distance of 400 m [43]. Several authors [42, 45, 64] indicate that the wind direction affects the gene flow, which is ascribed to its influence on the flight of bees.

The main consequence of gene flow between crop and their wild relatives is the increasing of wild relative fitness as a consequence of introgressed genes, which can lead to the development of invasive weeds. Some studies confirmed fitness increase of hybrids between sunflower crop and their relatives [47], while the other [77] confirmed hybrids in the first generation after crossing had lower fitness than wild parent in natural habitats, but in the following generations, fitness of hybrid was recovered. Also, hybrids between crop and wild populations of sunflower express lower fertility than their wild counterparts [75]. Although, crop hybridization can reduce dormancy in a wild species, hybridization IMI-resistant hybrid and wild sunflower in Argentina did not alter seed dormancy [41], while F1 germination was greater in wild sunflower populations [126].

Strategies for prevention or reduction of gene flow between crop sunflower and its relatives can be developed based on understanding seed and pollen dispersal and influence of different factors on that processes. The biological barriers based on cytoplasmic male sterility, which disable of plants to produce viable pollen, could be good option to reduce gene flow in sunflower.

http://dx.doi.org/10.5772/67645

## Acknowledgements

This work was supported by Ministry of Educattion, Science and Technological Development of Serbia.

## Author details

Sava Vrbničanin[1]*, Dragana Božić[1] and Danijela Pavlović[2]

*Address all correspondence to: sava@agrif.bg.ac.rs

1 Faculty of Agriculture, University of Belgrade, Belgrade, Serbia

2 Institute for Plant Protection and Environment, Belgrade, Serbia

## References

[1] Jocić S, Malidža G, Hladni N, Gvozdenović S. Novi hibridi suncokreta tolerantni na tribenuron metil. Zbornik radova Naučnog instituta za ratarstvo i povrtarstvo 2008;**XXII**:45.

[2] Shepherd DN, Mangwende T, Martin DP, Bezuidenhout M, Kloppers FJ, Carolissen CH, Monjane AL, Rybicki EP, Thomson JA. Maize streak virus-resistant transgenic maize: a first for Africa. Plant Biotechnology Journal 2007;**5** 759-767.

[3] Tang W, Newton RJ, Lin JX, Charles TM. Expression of a transcription factor from *Capsicum annuum* in pine calli counteracts the inhibitory effects of salt stress on adventitious shoot formation. Molecular Genetics and Genomics 2006;**276** 242-253.

[4] Ye XD, Al-Babili S, Kloti A, Zhang J, Lucca P, Beyer P, Potrykus I. Engineering the provitamin A (beta-carotene) biosynthetic pathway into (carotenoid-free) rice endosperm. Science 2000;**287** 3003-3005.

[5] Duke SO. Why have no new herbicide modes of action appeared in recent years? Pest Management Science 2012;**68** 505-512.

[6] Green JM. Current state of herbicides in herbicide-resistant crops. Pest Management Science 2014;**70** 1351-1357.

[7] Ryan GF. Resistance of common groundsel to simazine and atrazine. Weed Science 1970; **18** 614-616.

[8] Mallory-Smith CA, Thill DC, Dial MJ. Identification of sulfonylurea herbicide-resistant prickly lettuce (*Lactuca serriola*). Weed Technology 1990;**4** 163-168.

[9] Hall JC, Donnelly-Vanderloo MJ, Hume DJ. Triazine resistant crops: the agronomic impact and physiological consequences of chloroplast mutation. In: Duke, SO. (ed.)

Herbicide-resistant crops: agricultural, environmental, economic, regulator, and technical aspects. Florida: CRC Press, Boca Raton; 1996. pp. 107-126.

[10] Duke SO. Taking stock of herbicide-resistant crops ten years after introduction. Pest Management Science 2005;**61** 211-218.

[11] Stalker DM, Kiser JA, Baldwin G, Coulombe B, Houck CM. Cotton weed control using the BXN system. In: Duke SO, (ed.) Herbicide resistant crops: agricultural, environmental, economic, regulator, and technical aspects. Florida: CRC Press, Boca Raton; 1996. pp. 93-105.

[12] Duke SO. Perspectives on transgenic, herbicide-resistant crops in the United States almost 20 years after introduction. Pest Management Science 2015;**71** 652-657.

[13] Chaleff RS, Ray BT. Herbicide-resistant mutants from tobacco cell cultures. Science 1984;**223**(4641) 1148-1151.

[14] Newhouse K, Singh B, Shaner D, Stidham M. Mutations in corn (*Zea mays* L.) conferring resistance to imidazolinone herbicides. Theoretical and Applied Genetics 1991;**83** 65-70.

[15] Sebastian SA, Fader GM, Ulrich JF, Forney DR, Chaleff RS. Semi dominant soybean mutation for resistance to sulfonylurea herbicides. Crop Science 1989;**29** 1403-1408.

[16] Tan S, Evans RR, Dahmer ML, Singh BK, Shaner DL. Imidazolinone tolerant crops: history, current status and future. Pest Management Science 2005;**61** 246-257.

[17] Canadian Food Inspection Agency. Determination of the Safety of Pioneer Hi-Bred Production Ltd.'s Sulfonylurea-Tolerant ExpressSunTM Sunflower (*Helianthus annuus* L.) SU7. Decision Document DD2008-69. http://www.inspection.gc.ca/english/plaveg/bio/dd/dd0869e.shtml (accessed 1 March 2012).

[18] Cantamutto M, Poverene M. Genetically modified sunflower release: opportunities and risks. Field Crops Research 2007;**101**:133-144.

[19] James C. Global Status of Commercialized Biotech/GM Crops: 2005. International Service for the Acquisition of Agri-biotech Applications (ISAAA) Brief 34. http://www.isaaa.org

[20] Green MJ. The benefits of herbicide-resistant crops. Pest Management Science 2012;**68**: 1323-1331.

[21] Gressel J, Gassmann AJ, Owen MD. How well will stacked transgenic pest/herbicide resistances delay pests from evolving resistance? Pest Management Science 2017;**73**:22-34.

[22] GM Crop Database. Center for Environmental Risk Assessment (CERA). http://www.cera-gmc.org (accessed 12 January 2017).

[23] Gressel J. Crops with target-site herbicide resistance for *Orobanche* and *Striga* control. Pest Management Science 2009;**65**:560-565.

[24] Carpenter JE. Peer-reviewed surveys indicate positive impact of commercialized GM crops. Nature Biotechnology 2010;**28**:319-321.

[25] Malidza G, Jocic S, Skoric D, Orbovic B. Clearfield production system for sunflower (In Serbian). Proceedings of the Institute of Field and Vegetable Crops 2004;**40**:279-290.

[26] Schmidt LA, Talbert RE, McClelland M. Management of acetolactate synthase (ALS)-resistant common cocklebur (*Xanthium strumarium*) in soybean. Weed Technology 2004;**18**:665-674.

[27] Gianessi LP. Economic and herbicide use impacts of glyphosate-resistant crops. Pest Management Science 2005;**61**:241-245.

[28] Hurley TM, Mitchell PD, Frisvold GB. Characteristics of herbicides and weed-management programs most important to corn, cotton, and soybean growers. AgBioForum 2009;**12**:269-280.

[29] Oliveira AR, Castro TR, Capalbo DMF, Delalibera I. Toxicological evaluation of genetically modified cotton (Bollgard (R)) and Dipel (R) WP on the non-target soil mite Scheloribates praeincisus (Acari: Oribatida). Experimental and Applied Acarology 2007;**41**:191-201.

[30] Ellstrand NC, Prentice HC, Hancock JF. Gene flow and introgression from domesticated plants into their wild relatives. Annual Review of Ecology and Systematics 1999;**30**:539-563.

[31] Lu BR, Snow AA. Gene flow from genetically modified rice and its environmental consequences. Bioscience 2005;**55**:669-678.

[32] Kumar V, Bellinder RR, Brainard CD, Malik KR, Gupta KR. Risks of herbicide-resistant rice in India: A review. Crop Protection 2008;**27**:320-329.

[33] York AC, Steward AM, Vidrine PR, Culpepper AS. Control of volunteer glyphosate-resistant cotton in glyphosate resistant soybean. Weed Technology 2004;**18**:532-539.

[34] Rainbolt CR, Thill DC, Young FL. Control of volunteer herbicide-resistant wheat and canola. Weed Technology 2004;**18**:711-718.

[35] Messéan A, Sausse C, Gasquez J, Darmency H. Occurrence of genetically modified oilseed rape seeds in the harvests of subsequent conventional oilseed rape over time. European Journal of Agronomy 2007;**27**:115-122.

[36] Heap I. The International Survey of Herbicide Resistant Weeds. http://www.weedscience.org (accessed 13 January 2017).

[37] Beckie HJ, Hall LM. Simple to complex: modeling crop pollen mediated gene flow. Plant Science 2008;**175**:615-628.

[38] Lu BR, Yang C. Gene flow from genetically modified rice to its wild relatives: assessing potential ecological consequences. Biotechnology Advances 2009;**27**:1083-1091.

[39] Perez-Jones A, Martins B, Mallory-Smith CA. Hybridization in a commercial production field between imidazolinone-resistant wheat (*Triticum aestivum*) and *Aegilops cylindrica* results in pollen-mediated gene flow of Imi1. Weed Science 21;**58**:395-401.

[40] Arrigo N, Guadagnuolo R, Lappe S, Pasche S, Parisod C, Felber F. Gene flow between wheat and wild relatives: Empirical evidence from *Aegilops geniculata, Ae. neglecta* and *Ae. triuncialis*. Evolutionary Applications 2011;**4**:685-695.

[41] Presotto A, Ureta MS, Cantamutto M, Poverene M. Effects of gene flow from IMI resistant sunflower crop to wild *Helianthus annuus* populations. Agriculture, Ecosystems and Environment 2012;**146**:153-161.

[42] Bozic D, Pavlovic D, Bregola V, Di Loreto A, Bosi S, Vrbnicanin S. Gene flow from herbicide-resistant sunflower hybrids to weedy sunflower. Journal of Plant Diseases and Protection 2015;**122**:183-188.

[43] Whitton D, Wolf E, Arias DM, Snow AA, Rieseberg LH. The persistence of cultivar alleles in wild populations of sunflowers five generations after hybridization. Theoretical and Applied Genetics 1997;**95**:33-40.

[44] Lu CM, Kato M, Kakihara F. Destiny of a transgene escaped from *Brassica napus* into *B. rapa*. Theoretical and Applied Genetics 2002;**105**:78-84.

[45] Marshall MW, Al-Khatib K, Loughin T. Gene flow, growth, and competitiveness of imazethapy-resistant common sunflower. Weed Science 2001;**49**:14-21.

[46] Mercer KL, Wyse DL, Shaw RG. Effects of competition on fitness of wild and crop-wild hybrid sunflower from a diversity of wild populations and crop lines. Evolution 2006;**60**:2044-2055.

[47] Mercer KL, Andow DA, Wyse DL, Shaw RG. Stress and domestification traits increase the relative fitness of crop-wild hybrids in sunflower. Ecology Letters 2007;**10**:383-393.

[48] Campbell LG, Snow AA. Competition alters life history and increases the relative fecundity of crop-wild radish hybrids (*Raphanus* spp.). New Phytologist 2007;**173**:648-660.

[49] Loureiro I, Escorial MC, García-Baudín JM, Chueca MC. Evidence of natural hybridization between Aegilops geniculata and wheat under field conditions in Central Spain. Environmental Biosafety Research 2006;**5**:105-109.

[50] Zaharieva M, Monneveux P. Spontaneous hybridization between bread wheat (*Triticum aestivum* L.) and its wild relatives in Europe. Crop Science 2006;**46**:512-527.

[51] Tranel PJ, Wright TR. Resistance of weeds to ALS-inhibiting herbicides: what have we learned? Weed Science 2002;**50**:700-712.

[52] Molin WT, Nandula VK, Wright AA, Bond JA. Transfer and expression of ALS inhibitor resistance from Palmer Amaranth (*Amaranthus palmeri*) to *A. spinosus* × *A. palmeri* hybrid. Weed Science 2016;**64**:240-247.

[53] Campbell LG, Snow AA, Ridley EC. Weed evolution after crop gene introgression: greater survival and fecundity of hybrids in a new environment. Ecology Letters 2006;**9**(11):1198-1209.

[54] Gressel J. Dealing with transgene flow of crop protection traits from crops to their relatives. Pest Management Science 2015;**71**:658-667.

[55] Snow AA, Uthus KL, Culley TM. Fitness of hybrids between weedy and cultivated radish: implications for weed evolution. Ecological Applications 2001;**11**(3):934-943.

[56] Hedge SG, Nason JD, Clegg JM, Ellstrand NC. The evolution of California's wild radish has resulted in the extinction of its progenitors. Evolution 2006;**60**:1187-1197.

[57] Ellstrand NC, Elam D. Population genetic consequences of small population size: implications for plant conservation. Annual Review of Ecology and Systematics 1993;**24**(1): 217-242.

[58] Mallory-Smith C, Zapiola M. Gene flow from glyphosate-resistant crops. Pest Management Science 2008;**64**:428-440.

[59] Squire GR. Contribution to gene flow by seed and pollen. In: Messean A, (ed.) Proceedings of the 2nd International Conference on the Coexistence of Genetically Modified and Non-GM Based Agricultural Supply Chains. Agropolis Productions, Montpellier, France; 2005. pp. 73-77.

[60] Zhang W, Linscombe SD, Webster E, Tan S, Oard J. Risk assessment of the transfer of imazethapyr herbixcide tolernce from Clearfield rice to red rice (*Oryza sativa*). Euphytica 2006;**152**:75-86.

[61] Zapiola ML, Campbell CK, Butler MD, Mallory-Smith CA. Escape and establishment of transgenic glyphosate-resistant creeping bentgrass *Agrostis stolonifera* in Oregon, USA: a 4-year study. Journal of Applied Ecology 2008;**45**:486-494.

[62] Jenczewski E, Ronfort J, Chevre AM. Crop-to-wild gene flow, introgression and possible fitness effects of transgenes. Environmental Bisafety Research 2003;**2**:9-24.

[63] Lu Y, Burgos NR, Wang W, Yu L. Transgene flow from glufosinate-resistant rice to improved and weedy rice in China. Rice Science 2014;**21**:271-281.

[64] Massinga RA, Al-Khatib K, St Amand P, Miller JF. Gene flow from imidazolinone-resistant domesticated sunflower to wild relatives. Weed Science 2003;**51**:854-862.

[65] Faure N, Serieys H, Bervillé A. Potential gene flow from cultivated sunflower to volunteer, wild Helianthus species in Europe. Agriculture, Ecosystems and Environment 2002;**89**:183-190.

[66] Hvarleva T, Hristova M, Bakalova A, Hristov M, Atanassov I, Atanassov A. CMS lines for evaluation of pollen flow in sunflower relevance for transgene flow mitigation. Biotechnology and Biotechnological Equipment 2009;**23**:1309-1315.

[67] Renno JF, Winkel T. Phenology and reproductive effort of cultivated and wild forms of *Pennisetum glaucum* under experimental conditions in the Sahel: implications for the maintenance of polymorphism in the species. Canadian Journal of Botany 1996;**74** (6):959-964.

[68] Gutierrez A, Carrera A, Basualdo J, Rodrigez R, Cantamutto M, Poverene M. Gene flow between cultivated sunflower and *Helianthus petiolaris* (Asteraceae). Euphytica 2010;**172**:67-76.

[69] Burke J, Gardner K, Rieseberg L. The potential for gene flow between cultivated and wild sunflower (*Helianthus annuus*) in the United States. American Journal of Botany 2002;**89**:1550-1552.

[70] Gotz R, Ammer F. Ergebnisse der Anwendung von Liberty in transgenem Winterraps in Thuringen. Journal of Plant Diseases and Protection 2000;**27**:397-401.

[71] Song ZP, Lu B, Zhu YG, Chen JK. Gene flow from cultivated rice to the wild specied *Oryza rufipogon* experimental field conditions. New Phytologist 2003;**157**:657-665.

[72] Chevre AM, Eber F, Darmency H, Fleury A, Picault H, Letanneur JC, Renard M. Asessment of interspecific hybridization between transgenic oilseed rape and wild radish under normal agronomic conditions. Theoretical and Applied Genetics 2000;**100**:1233-1239.

[73] Warwick SI, Simard MJ, Légère A, Beckie HJ, Braun L, Zhu B, Mason P, Séguin-Swartz G, Stewart CN Jr. Hybridization between transgenic *Brassica napus* L. and its wild relatives: *B. rapa* L., *Raphanus raphanistrum* L., *Sinapis arvensis* L., and *Erucastrum gallicum* (Willd.) O. E. Schulz. Theoretical and Applied Genetics 2003;**107**:528-539.

[74] Linder CR, Taha I, Seiler GJ, Snow AA, Rieseberg LH. Long-term introgression of crop genes into wild sunflower populations. Theoretical and Applied Genetics 1998;**96**:339-347.

[75] Cummings CL, Alexander HM, Snow AA, Rieseberg LH, Kim MJ, Culley TM. Fecundity selection in a sunflower crop-wild study: can ecological data predict crop allele changes? Ecological Applications 2002;**12**:1661-1671.

[76] Stewart CN, Halfhill MD, Warwick SI. Transgene introgression from genetically modified crops to their wild relatives. Nature Reviews Genetics 2003;**4**:806-817.

[77] Gutierrez A, Cantamutto M, Poverene M. Persistence of sunflower crop traits and fitness in *Helianthus petiolaris* populations. Plant Biology 2011;**13**:821-830.

[78] Arnaud JF, Viard F, Delescluse M, Cuguen J. Evidence for gene flow via seed dispersal from crop to wild relatives in *Beta vulgaris* (Chenopodiaceae): consequences for the release of genetically modified crop species with weedy lineages. Proceedings of the Royal Society, Biological Sciences 2003;**270**:1565-1571.

[79] Raybould AF. Transgenes and agriculture-going with the flow? Trends in Plant Science 1999;**4**:247-248.

[80] Desplanque B, Hautekeete NC, van Dijk H. Transgenic weed beets: possible, probable, avoidable? Journal of Applied Ecology 2002;**39**:561-571.

[81] Cerdeira AL, Duke SO. The current status and environmental impacts of glyphosate-resistant crops: a review. Journal of Environmental Quality 2006;**35**:1633-1658.

[82] Desplanque B, Boudry P, Broomberg K, Saumitou-Laprade P, Cuguen J, van Dijk H. Genetic diversity and gene flow between wild, cultivated and weedy forms of *Beta vulgaris* L. (Chenopodiaceae) assessed by RFLP and microsatellite markers. Theoretical and Applied Genetics 1999;**98**:1194-1201.

[83] Messeguer J, Marfa V, Catala MM, Guiderdoni E, Mele E. A field study of pollen-mediated gene flow from Mediterranean GM rice to conventional rice and the red rice weed. Molecular Breeding 2004;**13**:103-112.

[84] Raybould AF, Gray AJ. Genetically modified crops and hybridization with wild relatives: a UK perspective. Journal of Applied Ecology 1993;**30**:199-219.

[85] Hancock JF. A framework for assessing the risk of transgene crops. Bioscience 2003;**53**: 512-519.

[86] Božić D, Stojićević D, Saulić M, Vrbničanin S. Transfer gena odgovornih za tolerantnost na herbicide sa useva na divlje srodnike. Acta Herbologica 2014;**23**:63-75.

[87] Scheffler JA, Parkingston R, Dale PJ. Frequency and distance of pollen dispersal from transgenic oilseed rape (*Brassica napus*). Transgenic Research 1993;**2**:356-364.

[88] Scheffler JA, Parkingston R, Dale PJ. Evaluating the effectiveness of isolation distances for field plots of oilseed rape (*Brassica napus*) using a herbicide-resistance transgene as a selectable marker. Plant Breeding 1995;**114**:317-321.

[89] Arias DM, Rieseberg LH. Gene flow between cultivated and wild sunflowers. Theoretical and Applied Genetics 1994;**89**:655-660.

[90] Eastham K, Sweet J. Genetically Modified Organisms (GMOs): The Significance of Gene Flow Through Pollen Transfer. European Environment Agency, Copenhagen, Denmark; 2002.

[91] Jarosz N, Loubet B, Durand B, Foueillassar X, Huber L. Variations in maize pollen emission and deposition in relation to microclimate. Environmental Science and Technology 2005;**39**:4377-4384.

[92] Aylor DE, Schultes NP, Shields EJ. An aerobiological framework for assessing cross-pollination in maize. Agricultural and Forest Meteorology 2003;**119**:111-129.

[93] Sanvido O, Widmer F, Winzeler M, Streit B, Szerencsits E, Bigler F. Definition and feasibility of isolation distances for transgenic maize, Transgenic Research 2008;**17**:317-355.

[94] Devos Y, Demont M, Sanvido O. Coexistence in the EU-return of the moratorium on GM crops? Nature Biotechnology 2008;**26**:1223-1225.

[95] Della Porta G, Ederle D, Bucchini L, Prandi M, Verderio A, Pozzi C. Maize pollen mediated gene flow in the Po valley (Italy): source-recipient distance and effect of flowering time. European Journal of Agronomy 2008;**28**:255-265.

[96] Palaudelmas M, Mele E, Penas G, Pla M, Nadal A, Serra J, Salvia J, Messeguer J. Sowing and flowering delays can be an efficient strategy to improve coexistence of genetically modified and conventional maize. Crop Science 2008,**48**:2404-2413.

[97] Nieh SC, Lin WS, Hsu YH, Shieh GJ, Kuo BJ. The effect of flowering time and distance between pollen source and recipient on maize. GM Crops and Food 2014;**5**(4):287-295.

[98] Morris WK, Kareiva PM, Raymer PL. Do barren zones and pollen traps reduce gene escape from transgenic crops? Ecological Applications 1994;**4**:157-165.

[99] Staniland BK, McVetty PBE, Friesen LF, Yarrow S, Freyssinet G, Freyssinet M. Effectiveness of border areas in confining the spread of transgenic *Brassica napus* pollen. Canadian Journal of Plant Science 2000;**80**:521-526.

[100] Weider C, Camp KH, Christov N, Hüsken A, Fouiellassar X, Stamp P. GM pollen containment by cytoplasmic male sterility (CMS) in maize (*Zea mays* L.). In: Stein AJ, Rodríguez-Cerezo E, (eds.) Book of abstracts of the 3rd International Conference on Coexistence between Genetically Modified (GM) and non-GM-based Agricultural Supply Chains. European Commission; 2007. pp. 65-68.

[101] Daniell H, Datta R, Varma S, Gray S, Lee SB. Containment of herbicide resistance through genetic engineering of the chloroplast genome. Nature Biotechnology 1998;**16**:345-348.

[102] Ruf S, Hermann M, Berger IJ, Carrer H, Bock R. Stable genetic transformation of tomato plastids-high-level foreign protein expression in fruits. Nature Biotechnology 2001;**19**:870-875.

[103] Daniell H. Molecular strategies for gene containment in transgenic crops. Nature Biotechnology 2002;**20**:581-586.

[104] Malidza G, Vrbnicanin S, Bozic D, Jocic S. Integrated weed management in sunflower: challenges and opportunities. Prceedings of the 19th International Sunflower Conference. Edirne, Turkey; 2016. pp. 88-97.

[105] Sala CA, Bulos M, Echarte M, Whitt SR, Ascenzi R. Molecular and biochemical characterization of an induced mutation conferring imidazolinone resistance in sunflower. Theoretical and Applied Genetics 2008;**118**:105-112.

[106] Al-Khatib K, Baumgartner JR, Peterson DE, Currie RS. Imazethapyr resistance in common sunflower (*Helianthus annuus*). Weed Science 1998;**46**:403-407.

[107] Sala CA, Bulos M, Echarte AM, Whitt SR, Ascenzi R. Development of CLHA-Plus: A Novel Herbicide Tolerance Trait in Sunflower Conferring Superior Imidazolinone Tolerance and Ease of Breeding. Proceedings of XVII International Sunflower Conference, Cordoba, Espana; 2008. pp. 489-494.

[108] Vrbničanin S, Božić D, Malidža G, Dušanić N, Pavlović D, Barać M. Tolerance of sunflower (*Helianthus annuus* L.) to imazethapyr. Helia 2008;**31**:85-94.

[109] Božić D, Sarić M, Malidža G, Ritz C, Vrbničanin S. Resistance of sunflower hybrids to imazamox and tribenuron-methyl. Crop Protection 2012;**39**:1-10.

[110] Miller JF, Al-Khatib K. Registration of two oilseed sunflower genetic stocks, SURES-1 and SURES-2, resistant to tribenuron herbicide. Crop Science 2004;**44**:1037-1038.

[111] Brankov M, Božić D, Sarić M, Vrbničanin S. Response of hybrids SUMO 1 PR and Kazanova to tribenuron-methyl (In Serbian). Zbornik rezima X savetovanja o zaštiti bilja; 2010. p. 76.

[112] Stanković-Kalezić R, Kojić M, Vrbničanin S, Radivojević Lj. *Helianthus annuus*—a new important element of the non-arable and arable flora in Serbia's region of Southern Banat. Helia 2007;**30**:37-42.

[113] Saulić M, Stojićević D, Matković A, Božić D, Vrbničanin S. Population variability of weedy sunflower as invasive species. Abstract book 4th ESENIAS Workshop: International Workshop on IAS in Agricultural and Non-Agricultural Areas in ESENIAS Region, Çanakkale, Turkey; 2013. p. 27.

[114] Muller MH, Latreille M, Tollon C. The origin and evolution of a recent agricultural weed: population genetic diversity of weedy populations of sunflower (*Helianthus annuus* L.) in Spain and France. Evolutionary Applications 2011;**4**:499-514.

[115] Trifković M, Saulić M, Stojićević D, Božić D, Vrbničanin S. Population variability and generative production of weedy sunflower (*Helianthus annuus* L.) (In Serbian). Zbornik rezimea XII savetovanje o zaštiti bilja. Zlatibor; 2013. p. 46.

[116] Stojicevic D, Saulic M, Dimitrijevic A, Bozic D, Miladinovic D, Vrbnicanin S. Population variability of weedy sunflower based on seed morphological traits. Book of Abstracts of the 4th International Symposium-Environmental Weeds and Invasive Plants. Monpellier, France; 2014. p. 105.

[117] Vrbničanin S, Božić D, Pavlović D, Sarić M. Fitness of the populations of invasive volunteer sunflower. Book of Abstracts 2nd International Workshop on Invasive Plants in the Mediterranean Type Regions of the World. Trabzon-Turkey, 2010; p. 85.

[118] Vrbničanin S, Božić D, Sarić-Krsmanović M, Stojićević D, Onć-Jovanović E. Fitnees of weedy sunflower populations (*Helianthus annuus* L.). In Serbian. Acta Herbologica 2013;**22**(2):89-99.

[119] Božić D, Sarić M, Elezović I, Vrbničanin S. Response of *Xanthium strumarium* L. and *Helianthus annuus* L. population to nicosulfuron (in Serbian). Acta Herbologica 2011;**20**:15-24.

[120] Božić D, Vrbničanin C, Barać M, Alić-Krsmanović M. ALS enzyme activity of *Xanthium strumarium* L. and *Helianthus annuus* L. influenced by nicosulfuron. Acta Herbologica 2013;**22**:79-87.

[121] Stojicevic D, Bozic D, Saulic M, Savic A, Vrbnicanin S, Petrovic I, Grujic M. Gene flow between different forms of sunflower (*Helianthus annuus* L.). Book of Abstracts of the 7th International Weed Science Congress. Prague, Czech Republic; 2016. p. 477.

[122] Gandhi SD, Heesacker AF, Freeman CA, Argyris J, Bradford K, Knapp SJ. The self-incompatibility locus (S) and quantitative trait loci for self-pollination and seed dormancy in sunflower. Theoretical Applied Genetics 2005;**111**:619-629

[123] Astiz V, Iriarte LA, Flemmer A, Hernandez LF. Self compatibility in modern hybrids of sunflower (*Helianthus annuus* L.) fruit set in open and self-pollinated (bag isolated) plants grown in two different locations. Helia 2011;**34**:129-138.

[124] Reagon M, Snow AA. Cultivated *Helianthus annuus* (Asteraceae) volunteers as a genetic "bridge" to weedy sunflower populations in North America. American Journal of Botany 2006;**93**(1):127-133.

[125] Rieseberg LH, Whitton J, Gardner K. Hybrid zones and genetic architecture of barrier to gene flow between two sunflower species. Genetics 1999;**152**:713-727.

[126] Mercer KL, Shaw RG, Wyse DL. Increased germination of diverse crop hybrid sunflower seeds. Ecological Applications 2006b;**16**:845-854.

Chapter 4

# Using Herbicide Programs to Control Weeds in Corn (*Zea mays* L.) and Cotton (*Gossypium hirsutum* L.)

W. James Grichar, Joshua A. McGinty,
Peter A. Dotray and Travis W. Janak

Additional information is available at the end of the chapter

http://dx.doi.org/10.5772/intechopen.68226

**Abstract**

Field studies were conducted to evaluate control of Amaranthus species and other weeds in corn and cotton. In corn, Palmer amaranth control was at least 90% with preemergence applications of fluthiacet-methyl plus pyroxasulfone, atrazine plus either acetochlor, alachlor, dimethenamid-P, *S*-metolachlor, or *S*-metolachlor plus mesotrione, saflufenacil plus dimethenamid-P, and *S*-metolachlor plus mesotrione. When using postemergence herbicides applied to Palmer amaranth less than 5 cm tall, atrazine, prosulfuron, and topramezone alone or the combinations of atrazine plus *S*-metolachlor plus glyphosate, diflufenzopyr plus dicamba, dimethenamid plus glyphosate, halosulfuron-methyl plus dicamba, mesotrione plus *S*-metolachlor plus glyphosate, pyroxasulfone plus glyphosate, and thiencarbazone-methyl plus tembotrione provided at least 91% control. In cotton, pyrithiobac applied preemergence resulted in no greater than 63% of control of Palmer amaranth and common waterhemp at the early season rating. Pendimethalin applied preemergence provided varied levels of control of common waterhemp. Trifluralin, applied preplant incorporated, consistently provided at least 86% or greater control of both species. A decreased level of control of both Palmer amaranth and common waterhemp was observed with pendimethalin applied preemergence followed by pyrithiobac-applied early postemergence and followed by glufosinate applied mid-post. Systems which included an early postemergence and mid-postemergence application of glyphosate plus 2,4-D choline provided at least 94% season-long Palmer amaranth control.

**Keywords:** annual grasses, broadleaf weeds, weed efficacy, crop response, *Amaranthus palmeri* S. Wats, *Amaranthus rudis* Sauer

## 1. Introduction

During the past 20 years, the use of glyphosate-resistant crop production systems has been adopted and used extensively in various regions of the USA [1]. In 2009, nearly 61 million ha of soybean [*Glycine max* (L.) Merr.], cotton (*Gossypium hirsutum* L), and corn (*Zea mays* L.) contained a modified 5-enolpyruvylshikimate-3-phosphate synthase (EPSPS) gene that confers resistance to glyphosate [2]. The wide use of row crops with glyphosate-resistance, the reduction of traditional herbicide and cultivation practices, and the use of intense management of weeds using glyphosate as the predominant control strategy has caused a shift in weed populations and created a selective advantage for glyphosate-resistant weeds [3, 4].

The development of herbicide-resistant crops allows weed control by nonselective postemergence (POST) herbicides, such as glyphosate and glufosinate, widening the array of weed management programs available to producers [5–7]. Both glyphosate and glufosinate control a wide range of weeds in herbicide-resistant crops [7] with little, if any, crop injury [8, 9]. POST applications of glyphosate or glufosinate provide consistent and greater control of large-seeded broadleaf weed species including velvetleaf (*Abutilon theophrasti* Medik.), giant ragweed (*Ambrosia trifida* L.), common cocklebur (*Xanthium strumarium* L.), and morningglory spp. (*Ipomoea* spp.) compared with preemergence (PRE) herbicides [9]. Even though the performance of glyphosate and glufosinate is similar, glufosinate is less likely to succeed in a single POST application program since glufosinate is less effective on larger weeds, needs an increased spray volume, and a need for high humidity at application [7].

Glyphosate-resistant weeds, specifically Amaranthus species, have become an issue across all the USA corn and cotton-producing areas [10]. Estimates are that more than 1.2 million ha of cropland in the USA are now affected by glyphosate-resistant Amaranthus species [10]. In cotton, Palmer amaranth (*Amaranthus palmeri* S. Wats.) has been shown to reduce lint yield by 57% when growing at a density of 10 plants per 9.1 m of row [11]. Additionally, with Palmer amaranth growing at densities greater than six plants per 9.1 m of row, cotton may not be harvestable due to the potential for damage to harvest equipment [11]. A study by Smith et al. [12] found that Palmer amaranth densities of 650–3260 plants $ha^{-1}$ in dryland stripper-harvested cotton increased harvesting time by 2- to 3.5-fold.

Weed resistance to photosystem II (PSII)-inhibiting herbicides, such as atrazine, has also been documented across many corn-growing areas of the USA [10]. Resistance to PSII inhibitors has been documented in 7 monocot and 17 dicot species in the corn-producing regions [13]. Also, populations of tall waterhemp [*Amaranthus tuberculatus* (Moq.) Sauer] have been identified with resistance to herbicides that inhibit acetolactate synthase (ALS), PSII, protoporphyrinogen oxidase (PPO), 5-enolpyruvylshikimate-3-phosphate-synthase (EPSPS), and 4-hydroxyphenyl-pyruvate-dioxygenase (HPPD) in Illinois and Iowa, and Palmer amaranth populations resistant to ALS, PSII, and HPPD inhibitors have been identified in Kansas [13], indicating the continued need for alternative modes of action in corn to reduce the chance of herbicide resistance. The HPPD-inhibiting herbicides have become popular among corn producers because of their broad-spectrum weed control, flexible application timings, tank-mix compatibilities, and crop safety [14–16].

Cotton growers have experienced more problems with weed resistance because of cotton's slower emergence after planting and fewer registered herbicides compared with other major crops [17]. The first documented cases of glyphosate-resistant (GR) Palmer amaranth in cotton occurred in 2000 in Lauderdale County, TN [18] and in 2003 in Edgecombe County, NC [19]. The first confirmed case of GR Palmer amaranth was documented in a biotype of Palmer amaranth growing in a Macon County, GA cotton field, where six- to eightfold levels of resistance to glyphosate were observed [3].

With the widespread adoption of glyphosate-resistant cotton after its introduction in 1997, cotton weed management practices largely shifted away from the use of soil-applied residual herbicides to POST herbicide programs based on glyphosate [20]. Studies conducted in 2006 and 2007 by Legleiter and Bradley [21] confirmed glyphosate resistance in a biotype of common waterhemp (*Amaranthus rudis* Sauer) found in a Missouri soybean field following multiple glyphosate applications. Currently, glyphosate-resistant Palmer amaranth and common waterhemp have been reported in 27 and 18 USA states, respectively [10]. Through surveys sent to weed scientists across the USA, Culpepper [3] revealed that 50% of respondents indicated that weeds of the genus Amaranthus had increased significantly in cotton. The respondents also provided the following four recommendations for managing glyphosate-induced weed species shifts: tank-mix combinations of other herbicides with glyphosate for POST applications, rotating with non-glyphosate-resistant crops (though there was some disagreement among respondents), use of POST herbicides other than glyphosate, and using preplant-incorporated (PPI) or (PRE) soil-applied herbicides.

Amaranthus species are some of the most common weed species found in annual crop production throughout the USA [22]. Palmer amaranth is now ranked as the most troublesome weed found in the USA [23]. It is a common weed in many major crops around the world and is found in all areas of Texas [24]. Up until the 1990s, its distribution in North America was the southern half of the USA [24]; however, since then, it has become established in every state with the exception of the northwestern USA, including Washington, Oregon, Montana, and North Dakota [25]. In Texas, Palmer amaranth can be found in all areas of the state [26] and is one of the two Amaranthus species with confirmed resistant to glyphosate across Texas (common waterhemp is the other) [27]. It is a dioecious, summer-annual species that is native to the desert southwest region of the USA [28, 29]. Plants of the genus Amaranthus are often very problematic weeds in agronomic crops due to their ability to germinate under a wide range of conditions, grow rapidly, and produce large numbers of seed, all while competing with the crop for sunlight, moisture, and nutrients. Despite its origin, Palmer amaranth is able to survive in many diverse environments because of its biological characteristics [6, 30]. It has a lengthy germination window, robust growth habit, and is a prolific seed producer [31–33], and these characteristics make control of this weed difficult. Common waterhemp is an obligate outcrossing annual broadleaf weed that is capable of long-distance pollen dispersal [34]. It germinates optimally between 20/25 and 30/35°C [35], has an aggressive growth habit and may grow 1.6 mm per growing degree day [32], and is capable of producing more than 250,000 seeds per plant [30]. These factors make it a strong competitor with most crops.

Traditional corn and cotton weed management programs have relied on PRE applications of a broadleaf and grass herbicide for residual season-long weed control [36–41]. In corn, these PRE programs usually have included atrazine in combination for broad-spectrum weed control. Atrazine is used in over 60% of the USA corn, and its doses have gotten lower with most doses of no more than 1.12 kg $ha^{-1}$ with some growers applying no more than 0.84 kg $ha^{-1}$ [42]. Atrazine and 4-hydroxylphenylpyruvate dioxygenase (HPPD)-inhibiting herbicides are commonly used for weed control in corn and are effective in controlling glyphosate-resistant weeds, including Palmer amaranth [36, 43, 44]. Atrazine can be applied PRE or POST alone or in tank-mixtures with several herbicides [16].

Since POST herbicides are applied after the weed species and severity are known, this allows growers to assess the problem before making a herbicide application; therefore, POST herbicides are an essential component of an integrated weed management system to combat the herbicide-resistant weeds [45]. In addition, POST herbicides typically do not require rainfall for herbicide activation, making performance less dependent on environmental conditions [45]. Also, POST herbicides can reduce the potential for water pollution [46]. A Minnesota study showed reduced atrazine concentrations in runoff water when applied POST compared with soil-applied applications because of the increased plant residue and cover, limiting the amount of herbicide reaching the soil [47].

Two new herbicide systems have recently become important in POST weed control in cotton [48–53]. Dicamba (3,6-dichloro-2-methoxybenzoic acid) is synthetic auxin herbicide that controls glyphosate-resistant Palmer amaranth and other broadleaf weeds alone or in sequential combinations with glyphosate or glufosinate [48]. An enzyme, dicamba O-demethylase, was discovered in a soil bacterium (*Pseudomonas maltophilia*) that converts dicamba to 3,6-dichlorosalicylic acid (DCSA) [49]. The enzyme DCSA has no significant herbicidal properties. The gene responsible for this enzyme is known as DMO (dicamba monooxygenase). This gene was successfully inserted into mouse-ear cress [*Arabidopsis thaliana (L). Heynh.*], tomato (*Solanum lycopersicum* L.), and tobacco (*Nicotiana tabacum* L.) and showed to provide these plants with effective tolerance to foliar applications of dicamba [49]. Dicamba-tolerant cotton, coupled with existing glyphosate- and glufosinate-tolerant traits, was deregulated in the USA in 2015 and has since become significant portion of the cotton planted in the USA, comprising over 40% of the crop planted in 2016 [50, 51].

Enlist Duo herbicide, a premix formulation containing 195 g ae $L^{-1}$ of 2,4-D choline and 205 g ae $L^{-1}$ of glyphosate dimethylamine, was developed for use in Enlist corn, cotton, and soybean. Resistance to 2,4-D is conferred by the insertion of a gene that codes for the enzyme aryloxyalkanoate dioxygenase. Plants transformed to include this gene can metabolize 2,4-D to a nonlethal form [52]. Developed during World War II, 2,4-D was the first selective herbicide widely used in agriculture [53]. Since that time, researchers have demonstrated control of a large number of dicotyledonous weed species with 2,4-D [54–57].

The adoption of 2,4-D in Enlist crops will be influenced by yield potential of the crop, weed species infesting fields, and, most notably, the ability of growers to mitigate off-target movement of 2,4-D [58–60]. Although Enlist cotton is resistant to 2,4-D [61], all other cotton cultivars, including cotton resistant to dicamba, are extremely sensitive to the herbicide, with reports of cotton injury due to 2,4-D drift dating back to the time of development [62]. Multiple studies showed that exposure to 2,4-D resulted in cotton injury with sensitivity increasing at earlier growth stages and higher herbicide concentrations [63–65].

The prime strategy for managing herbicide resistance in weeds is to reduce the selection pressure for resistance evolution by any one selecting agent, while managing adequate weed control [66]. Selection pressure has the greatest impact on herbicide-resistance evolution and is a factor that growers can control. Selection pressure imposed by an herbicide is the product of efficacy and persistence in the soil [67]. Herbicides applied in crop generally result in the greatest selection pressure compared with other application timings. Selection pressure against a weed population over time, resulting in increasing frequency of resistant individuals that collectively possess one or more resistance mechanisms, is a function of frequency of application [66]. Herbicide sequences, rotations, or mixtures generally have the greatest effect in delaying resistance when the mechanism conferring resistance is target-based, the weed species are highly self-pollinated, and seed spread is restricted [68, 69] and herbicide mixtures may delay resistance longer than rotations [70].

The rapid increase in resistant weeds in corn and cotton and the concerns pertaining to the overuse of atrazine in corn, including detection in surface and groundwater, rotational crop injury, and the development of triazine-resistant weeds, calls for the development of appropriate and effective management techniques. Also, growing questions about the renewed use of PRE and POST herbicides for early season and possibly season-long weed control in corn and cotton have also become a major topic of discussion. Therefore, the objective of this research was to evaluate the effect of various PRE and POST herbicides alone and in combinations for crop tolerance and weed control efficacy in the Texas corn and cotton-producing regions. In cotton, several herbicide programs in glyphosate-, glufosinate-, and dicamba-tolerant cotton were evaluated for their efficacy on both Palmer amaranth and common waterhemp.

## 2. Materials and methods

### 2.1. Corn PRE studies in central and south Texas

These studies were conducted during the 2013 through 2015 growing season in central Texas near Taylor (30.5326° N, 97.4548° W) and in south-central Texas near Ganado (29.0438° N, 96.4849° W). Study sites were located in different fields within the same general area of each year. Soils at the Taylor location were a Burleson clay (fine, montmorillonitic, and thermic Udic Pellusterts) with less than 1% organic matter and 7.6 pH, while soils at the Ganado location were a Houston Black clay (fine, montmorillonitic, and thermic Udic Pellusterts) with less than 1% organic matter and 7.4 pH.

Studies were arranged in a randomized complete block design with three replicates. Plot dimensions were two or four corn rows, wide spaced 76–97 cm apart, and 6.3 or 7.9 m long (depending on location). The corn hybrids BH 8846RR (2013), BH 8844VTTP (2014), and BH 8475SS (2015) were planted mid- to late February near Taylor and late February to early March near Ganado in each year to a depth of 2.5–3.5 cm at the rate of 54,000–65,500 seeds $ha^{-1}$.

Herbicides were applied within 5–7 days after planting with a $CO_2$-pressurized backpack sprayer with TeeJet 11002 flat-fan nozzles (Spraying Systems Co., North Avenue and Schmale Road, Wheaton, IL 60188) using a pressure of 180 kPa and calibrated to deliver 140 or 187 L $ha^{-1}$ (depending on location). An untreated check was included for comparison at each location. All herbicide doses were based on the USA label dose with the exception of the acetochlor (74.8% formulation) dose which was applied at 2X of the labeled rate throughout the study by mistake. Once the error was realized, it was decided to maintain this dose throughout the study.

Weed populations varied from year to year and were from naturally occurring soil seed bank populations. At the Taylor location, browntop panicum [*Panicum fasciculatum* Sw. var. *reticulatum* (Torr.) Beal] populations in 2013 were moderate (3–4 plants/$m^2$), while in 2014 populations were higher (6–8 plants/$m^2$). Common barnyardgrass [*Echinochloa crus-galli* (L.) P. Beauv.] populations in 2015 ranged from 4–8 plants/$m^2$. At Ganado, Texas millet [*Urochloa texana* (Buckley) R. Webster] populations ranged from 6–10 plants/$m^2$. Palmer amaranth populations varied from 4–8 plants/$m^2$ at the Taylor location to 2–10 plants/$m^2$ at the Ganado location. Hophornbeam copperleaf (*Acalypha ostryifolia* Riddell) populations at Taylor in both years were low to moderate (2–6 plants/$m^2$), while common sunflower (*Helianthus annuus* L.) populations ranged from 2–6 plants/$m^2$ depending on the year.

Crop injury and weed control were estimated visually on a scale of 0–100 (0 indicating no control or injury and 100 indicating complete control or plant death). Crop injury consisted of plant stunting and early season (30 days after herbicide application) and late season (95–140 days after application) crop injury was recorded. Late season weed control ratings (95–140 days after herbicide application) are presented for all weeds with the exception of Palmer amaranth control at Ganado in 2015 where populations of this weed were low (<4 plants/$m^2$) and somewhat inconsistent. Crop yield was determined by hand-harvesting 3.8 m of each plot, shelling the kernels from the corn ear, and weighing the kernels. Crop weights were adjusted to 15% moisture.

Visual estimates of weed control and corn injury were transformed to the arcsine square root prior to analysis of variance but are expressed in their original form for clarity because the transformation did not alter interpretation. Means were compared with Fisher's Protected LSD test at the 5% probability level [71]. The non-treated check was not included in the weed control analysis but was included in corn yield analysis.

### 2.2. Corn POST studies in central and south Texas

Field studies were conducted during the 2013 through 2015 growing season at two locations in central Texas including near Taylor (30.5326° N, 97.4548° W) and Beyersville (30.3036° N, 97.1947° W) and at three locations in south-central Texas near Kendleton (29.44786° N, 95.99961° W), Ganado (29.0438° N, 96.4849° W), and Yoakum (29.1827° N, 97.0929° W). Where study sites were similar over years, these studies were located in different fields within the same general area. Soils at the central Texas locations near Taylor were a Burleson clay (fine, montmorillonitic, and thermic Udic Pellusterts) with less than 1% organic matter and 7.6 pH, while soils at the Beyersville location soils were a Houston Black clay (fine, smectitic, and thermic Udic

http://dx.doi.org/10.5772/intechopen.68226

Haplusterts) with less than 3% organic matter and 7.8 pH. Soils at the south-central locations near Ganado were a Laewest clay (fine, montmorillonitic, and thermic Udic Pellusterts) with less than 1% organic matter and 7.4 pH, soils at Kendleton were a Bernard-Edna complex (fine, smectitic, and hyperthermic Oxyaquic Vertic Argiudolls) with less than 3% organic matter and 6.8 pH. Soils at the Yoakum location were a Cuero sandy clay loam (fine, loamy, mixed, superactive, and thermic Pachic Argiustolls) with less than 2% organic matter and 7.2 pH.

Studies were arranged in a randomized complete block design with three replicates of treatments. Plot dimensions were either two or four rows (depending on location), spaced 76–97 cm apart by 6.3–7.9 m long. The corn varieties BH 8846RR (2013), BH 8844 VTTP (2014), and BH 8475 SS (2015) were planted from mid-February to mid-March depending on locations and environmental conditions to a depth of approximately 2.5–3.5 cm at the rate of 54,000–65,500 seeds $ha^{-1}$.

Herbicides were applied POST with a $CO_2$-pressurized backpack sprayer using TeeJet 11002 flat-fan nozzles (Spraying Systems Co., North Avenue and Schmale Road, Wheaton, IL 60188) with a pressure of 180 kPa and calibrated to deliver 140–187 L $ha^{-1}$ (depending on location). An untreated check was included for comparison at each location. All herbicide doses were based on the USA label and included an adjuvant and either ammonium nitrate or sulfate per label requirements.

Weed populations varied from location to location and were from natural seed bank populations in the soil. At the Taylor location, browntop panicum populations in 2013 were sparse (3–4 plants/$m^2$), while Texas millet populations at Beasley were extremely dense (16–18 plants/$m^2$) and moderate at Beyersville (6–8 plants/$m^2$). Common barnyardgrass pressure at Taylor was low to moderate (4–8 plants/$m^2$). Palmer amaranth populations at the Yoakum and Ganado locations was dense (16–20 plants/$m^2$), while populations at the Taylor locations were low (4–6 plants/$m^2$). Pitted morningglory (*Ipomoea lacunose* L.), hophornbeam copperleaf, and Asiatic dayflower (*Commelina communis* L.) populations were low to moderate (4–8 plants/$m^2$). Approximately 50% of the Palmer amaranth population at the Ganado location was glyphosate resistant.

Weed size at the time of treatment varied by location. Browntop panicum was no greater than 15 cm tall when treated, while Texas millet and common barnyardgrass were less than 20 cm tall at the time of herbicide application. Palmer amaranth at the Yoakum location was less than 5 cm tall at herbicide application, while at Taylor weed size was less than 20 cm. However, at the Ganado location, Palmer amaranth height varied from 40 to 60 cm due to rains which prevented entry into the field in a timely manner. Pitted morningglory length ranged from 5 to 20 cm, while hophornbeam copperleaf and Asiatic dayflower were less than 20 cm in height at the time of treatment. Corn height varied from location to location but was typically in the V4–V8 stage.

Crop injury and weed control were visually estimated on a scale of 0–100 (0 indicating no control or injury and 100 indicating complete control or plant death). Mid- to late season weed control ratings (31–98 days after herbicide application) are presented for all weeds. Crop yield was determined by hand-harvesting 3.8 m of each plot, shelling the kernels from the corn ear, and weighing the kernels. Crop weights were adjusted to 15% moisture.

Visual estimates of weed control and corn injury were transformed to the arcsine square root prior to analysis of variance but are expressed in their original form for clarity because the transformation did not alter interpretation. Means were compared with Fisher's Protected LSD test at the 5% probability level [71]. The non-treated check was not included in the weed control analysis but was included in corn yield analysis.

### 2.3. Cotton studies in south-central Texas

Studies were conducted in Burleson County, TX (30.3257° N, 96.2615° W) at the Texas A&M AgriLife Research Farm in 2012 and 2013 to investigate management strategies for controlling Palmer amaranth and common waterhemp in cotton possessing glyphosate-, glufosinate-, and dicamba-tolerant transgenic traits. Studies were in the same general area in each year. Soils at this site are characterized as a Westwood silty clay loam (fine, silty, mixed, superactive, and thermic Udifluventic Haplustepts) with 2% organic matter and 8.1 pH. The experiment included 12 treatments arranged as a randomized complete block design with 4 replications. Plots were four rows wide and 9.1 m in length with 102 cm row spacing. Buffers 4.5 m wide were maintained between blocks to facilitate lateral movement of equipment.

This experiment was conducted on a furrow-irrigated field with large seed bank populations of both Palmer amaranth and common waterhemp. Both Palmer amaranth and common waterhemp were naturally occurring populations with 10–15 plants/m$^2$. In 2012, none of the Palmer amaranth or common waterhemp populations were glyphosate resistant, while in 2013 approximately 10% of the Palmer amaranth population was resistant; however, none of the waterhemp populations were resistant. Treatments included preplant-incorporated (PPI), PRE, and two POST application timings of an early POST (EPOST) and mid-POST (MPOST). Plots receiving PPI applications of trifluralin were subjected to two passes of a rolling cultivator immediately following application to thoroughly incorporate the herbicide into the soil. Preemergence herbicide applications included fomesafen, pendimethalin, prometryn, pyrithiobac, and *S*-metolachlor, while POST applications included acetochlor, dicamba, glufosinate, glyphosate, pyrithiobac, and trifloxysulfuron. Early postemergence applications in 2012 were made when weeds were approximately 12 cm tall and in 2013 when weeds were 10 cm in height, while MPOST treatments in 2012 were made when weeds were 25 cm tall and in 2013 when 15 cm in height. An untreated check was included in all studies.

For the 2012 experiment, PPI applications were made on May 7, cotton was planted on May 22, EPOST applications were made on June 21, and MPOST applications were made on July 4. In 2013, PPI applications were made on May 8, cotton was planted on May 9, EPOST applications were made on June 7, and MPOST applications were made on June 16. The cotton variety was an experimental dicamba-glyphosate tolerant entry from Monsanto. Herbicide applications were made with a $CO_2$-pressurized backpack sprayer calibrated to deliver 140 L ha$^{-1}$ total spray volume. Preplant-incorporated and PRE applications were made using TeeJet 11003 Drift Guard flat-fan nozzles, while EPOST and MPOST applications were made with TeeJet 110015 Turbo TeeJet Induction flat-fan nozzles (TeeJet Technologies, Wheaton, Illinois 60187).

Control of Palmer amaranth and common waterhemp was estimated visually at the time of the EPOST application, at the time of MPOST application, and 14 days after the MPOST application.

These observations are reported as early, mid, and late, respectively. Plots were managed throughout the season according to standard crop management practices for this region. The center two rows of all plots were mechanically harvested and seed cotton yields were recorded. Means were compared with Fisher's Protected LSD test at the 5% probability level [71].

### 2.4. Cotton studies in the High Plains of Texas

Field studies were conducted near New Deal (33.4413° N, 101.4358° W) and Halfway, TX (34.1881° N, 101.9522° W) during the 2015 and 2016 growing seasons to investigate management strategies for controlling Palmer amaranth in cotton possessing glyphosate-, glufosinate-, dicamba-, and 2,4-D choline-tolerant transgenic traits. Soils at the New Deal site are characterized as a Pullman clay loam (fine, mixed, and superactive thermic Torrertic Paleustolls) with less than 1% organic matter and 7.9 pH, while soils at New Deal are a Olton clay loam (fine, mixed, and thermic Aridic Paleustoll) with less than 1% organic matter and a 7.4 pH. These experiments were conducted under center pivot irrigation at Halfway and 102 cm spacing of sub-surface drip tape at New Deal with large populations of Palmer amaranth (8–10 plants/m$^2$). These studies were conducted as a randomized complete block design with four replications. Plots were four rows wide and 9.1–12.7 m in length, with 102 cm row spacing.

Treatments for the glyphosate plus 2,4-D choline study (Enlist Duo) included PPI treatments of trifluralin and EPOST and MPOST treatments of glyphosate, glufosinate, glyphosate plus 2,4-D choline, *S*-metolachlor, and 2,4-D choline salt. Plots receiving PPI applications of trifluralin were subjected to two passes of a rolling cultivator immediately following application. Postemergence applications were made when Palmer amaranth was 15 cm or less in height. This study was conducted in 2016. The cotton variety was an experimental from Dow AgroSciences (9330 Zionsville Rd, Indianapolis, IN 46268) and was planted on May 26 at a seeding rate of 13.1 seeds m$^{-1}$ of row.

Treatments for the glyphosate systems study included preplant applications of glyphosate plus either flumioxazin, fomesafen, the premix of rimsulfuron plus thifensulfuron-methyl, or diruron, PRE applications of either flumeturon, pyrithiobac, acetochlor, or flumeturon plus paraquat, EPOST applications of glyphosate alone or plus either acetochlor, *S*-metochlor, dimethenamid-P, or pyrithiobac, MPOST applications of glyphosate alone or plus acetochlor, dimethenamid-P, or S-metochlor, and LPOST treatments of diruron plus MSMA. Postemergence applications were made when Palmer amaranth was 10 cm or less in height. This study was conducted in 2015 and 2016. Fibermax 2322GL was planted in both years with the same seeding rate as in the previous study.

Herbicide applications were made with a $CO_2$-pressurized backpack sprayer calibrated to deliver 93–140 L ha$^{-1}$ total spray volume. PPI and PRE applications were made using TeeJet 11002 Drift Guard flat-fan nozzles, while EPOST and MPOST applications were made with TeeJet 110015 Turbo TeeJet Induction flat-fan nozzles. A Redball® tractor-mounted hooded sprayer (Willmar Fabrication, LLC; Willmar, MN 56201) was used for LPOST herbicide applications.

Control of Palmer amaranth was estimated visually as in previous studies. Plots were managed throughout the season according to standard crop management practices for this region.

The center two rows of all plots were mechanically harvested, and lint cotton yields were recorded. Means were compared with Fisher's Protected LSD test at the 5% probability level [71].

## 3. Results and discussion

Although the primary emphasis in this chapter is the discussion on controlling Palmer amaranth and, to some extent, common waterhemp which have become troublesome weeds in corn and cotton, other weed species will be discussed since they are/can become problematic weeds as well.

### 3.1. Corn PRE studies

Since not all treatments were included in each year of the study, no attempt was made to combine results over years or locations. Also, rainfall amounts varied from site to site and year to year affecting herbicide response (**Table 1**). Rainfall during the 7 days after the application of PRE herbicide treatments occurred at all locations with the exception of Ganado in 2013 and 2014 when no rainfall occurred. Rainfall between 8 and 14 days after the PRE application varied from no rainfall at Ganado in 2013 to 78.2 mm at Ganado in 2015 (**Table 1**). Rainfall 15–21 days after the PRE application was low at Taylor in 2013 and Ganado in 2015, and no rainfall occurred at the other sites.

With respect to annual grasses, browntop panicum and Texas millet were present in 2013 and 2014 at the Taylor and Ganado sites, respectively. Common barnyardgrass was present at Taylor only in 2015. Broadleaf weeds were present at the Taylor and Ganado locations. Palmer amaranth was present in 2013 and 2015, while hophornbeam copperleaf and common sunflower were present in 2013 and 2014. Although this chapter discusses the control of herbicide-resistant weeds, the control of other weeds will also be discussed since they are also a large part of the problem when providing effective weed control under normal growing conditions.

#### *3.1.1. Annual grass control*

Atrazine alone controlled common barnyardgrass 33%, while acetochlor (74.8%) or pendimethalin alone, acetochlor plus atrazine, *S*-metolachlor plus mesotrione, or *S*-metolachlor plus

| | 2013 | | 2014 | | 2015 | |
|---|---|---|---|---|---|---|
| | **Taylor** | **Ganado** | **Taylor** | **Ganado** | **Taylor** | **Ganado** |
| **Day** | **Mm** | | | | | |
| 1–7 | 29.5 | 0 | 2.8 | 0 | 7.4 | 3.3 |
| 8–14 | 6.6 | 0 | 0.5 | 18.6 | 65.6 | 78.2 |
| 15–21 | 7.3 | 0 | 0 | 0 | 0 | 3.3 |

**Table 1.** Rainfall amounts at test locations for 21 days following application of PRE herbicides.

atrazine plus mesotrione provided 90–97% control (**Table 2**). The dinitroaniline herbicides, such as pendimethalin, are registered for use in over 40 crops [72]. These herbicides usually provide excellent control of annual grasses [73–75].

In 2013, pendimethalin alone, alachlor plus atrazine, *S*-metolachlor plus mesotrione, or *S*- metolachlor plus atrazine plus mesotrione provided 96% or better browntop panicum control, while isoxaflutole, *S*-metolachlor, and pyroxasulfone alone, and *S*-metolachlor plus atrazine controlled this weed 80–88% (**Table 2**). In 2014, only the dose of acetochlor (74.8%) provided acceptable control (83%). The lack of effective control in 2014 can be attributed to greater plant populations at the test site in 2014 compared to 2013 and also the low rainfall amounts after the PRE application in 2014 (**Table 1**). Since many of the PRE herbicides can volatilize and photodecompose on the soil surface over time, these herbicides need to be mechanically incorporated or need rainfall or irrigation to move these herbicides into the weed seed zone [76–78], which explains the erratic control noted with these herbicides under the droughty conditions observed at Taylor in 2014.

| | | **Browntop panicum** | | **Texas millet** | | **Barnyardgrass** |
|---|---|---|---|---|---|---|
| | **Dose** | **2013** | **2014** | **2013** | **2014** | **2015** |
| **Treatment** | **Kg ai ha$^{-1}$** | **Taylor** | | **Ganado** | | **Taylor** |
| | | Days after treatment | | | | |
| | | 95 | 138 | 109 | 112 | 101 |
| | | % | | | | |
| Atrazine (A) | 1.1 | 33 | 3 | 23 | 0 | 33 |
| Fluthiacet-methyl (FM)+pyroxasulfone (P) | 0.006+0.2 | - | - | - | - | 58 |
| (FM)+(P)+(A) | 0.004+0.2+1.3 | - | - | - | - | 40 |
| *S*-metolachor (S) | 1.3 | 82 | 57 | 78 | 75 | 68 |
| Isoxaflutole | 0.05 | 80 | 38 | 94 | 80 | 67 |
| (S)+(A) | 1.4+1.8 | 85 | 53 | 86 | 83 | 63 |
| Alachlor+(A) | 2.5+1.5 | 99 | - | 89 | - | - |
| Mesotrione (M) | 0.1 | 37 | 8 | 53 | 88 | 55 |
| Thiencarbazone-methyl+isoxaflutole | 0.02+0.06 | 47 | 15 | 98 | 72 | 73 |
| Acetochlor+(A) | 2.1+1.3 | 72 | 0 | 98 | 86 | 90 |
| (S)+(A)+(M)+bicyclopyrone | 0.4+0.8+0.09+0.02 | - | - | - | - | 65 |
| Dimethenamid-P+(A) | 1.6+3.2 | 55 | 45 | 85 | 58 | 89 |
| Acetochlor (74.8%) | 6.9 | - | 83 | - | 73 | 97 |
| Rimsulfuron+(M) | 0.02+0.2 | 60 | 33 | 77 | 47 | - |
| Rimsulfuron+thifensulfuron-methyl | 0.02+0.02 | 74 | 10 | 60 | 61 | 40 |
| (S)+(A)+(M) | 1.5+1.5+0.2 | 98 | 44 | 83 | 73 | 92 |

| | | Browntop panicum | | Texas millet | | Barnyardgrass |
|---|---|---|---|---|---|---|
| | **Dose** | **2013** | **2014** | **2013** | **2014** | **2015** |
| **Treatment** | **Kg ai ha$^{-1}$** | **Taylor** | | **Ganado** | | **Taylor** |
| Dimethenamid-P | 0.8 | 78 | 53 | 55 | 63 | 73 |
| Pendimethalin | 1.6 | 96 | 52 | 86 | 99 | 97 |
| Saflufenacil | 0.05 | 69 | 7 | 81 | 23 | 33 |
| Saflufenacil+dimethenamid-P | 0.08+0.7 | 61 | 28 | 92 | 78 | 57 |
| Acetochlor (33%) | 1.7 | 75 | 10 | 78 | 96 | 63 |
| (S)+(M) | 2.8+0.3 | 98 | 65 | 67 | 95 | 96 |
| (P) | 0.1 | 88 | 37 | 75 | 99 | 42 |
| Untreated | - | 0 | 0 | 0 | 0 | 0 |
| LSD (0.05) | | 33 | 33 | 22 | 48 | 29 |

**Table 2.** Annual grass control in corn with PRE herbicides.

In 2013, isoxaflutole alone, thiencarbazone-methyl plus isoxaflutole, acetochlor plus atrazine, or saflufenacil plus dimethenamid-P controlled Texas millet at least 92% (**Table 2**). Pendimethalin or saflufenacil alone, atrazine plus either *S*-metolachlor, alachlor, or dimethenamid-P, and the three-way combination of *S*-metolachlor plus atrazine plus mesotrione provided 81–89% control. In 2014, acetochlor, pendimethalin, or pyroxasulfone alone or *S*-metolachlor plus mesotrione controlled this weed at least 95%, while isoxaflutole or mesotrione alone and atrazine plus either acetochlor or *S*-metolachlor controlled 83–89% (**Table 2**). In the two years, *S*-metolachlor alone provided 75–78% Texas millet control compared with 75–99% control with pyroxasulfone. Typically, *S*-metolachlor alone provides poor control of this weed [79, 80]. With high populations of Texas millet, Grichar et al. [79] reported less than 70% control with 1.7 and 3.4 kg ha$^{-1}$ of metolachlor in dryland peanut (*Arachis hypogaea* L.) and 25–76% control under irrigated conditions. Steele et al. [80] reported that pyroxasulfone, at a 10-fold lower use rate than *S*-metolachlor, controlled Texas millet 84–96%, while *S*-metolachlor provided 75–85% control when rated 9 weeks after treatment. They attributed the results to the longer residual activity of pyroxasulfone [81].

### *3.1.2. Broadleaf weed control*

At Taylor in 2013, under moderate weed pressure (4 plants m$^2$), all herbicides, with the exception of atrazine (73%), provided at least 97% Palmer amaranth control, while in 2015 under increased populations (8 plants m$^2$), atrazine controlled Palmer amaranth 79%, while isoxaflutole, mesotrione, or saflufenacil provided no better than 71% control (**Table 3**). All other herbicide treatments provided at least 96% control. At the Ganado location, in 2013 and 2015, control was more erratic than at the Taylor location. This may be due to the greater weed populations noted in 2013 (10 plants m$^2$) and variable populations in 2015. In 2013, either atrazine or isoxaflutole alone, acetochlor, alachlor, *S*-metolachlor, or dimethenamid-P plus atrazine, or the three-way combination of *S*-metolachlor plus atrazine plus mesotrione provided 97–100% control, while mesotrione, dimethenamid-P, or acetochlor (33%) alone and rimsulfuron plus mesotrione controlled this weed 61% or less (**Table 3**). In 2015, acetochlor (74.8%) alone, dimethenamid-P plus atrazine, fluthiacet-methyl plus pyroxasulfone, and saflufenacil plus dimethenamid-P controlled Palmer amaranth at least 95%, while isoxaflutole, mesotrione, *S*-metolachlor, and pendimethalin alone and rimsulfuron plus thifensulfuron-methyl controlled this weed less than 70%.

In previous research, mesotrione applied PRE controlled smooth pigweed (*Amaranthus hybridus* L.), but control of morningglory species (*Ipomoea* spp.) and common lambsquarter (*Chenopodium album* L.) was inconsistent and dependent upon a timely rainfall following application [38, 82]. Armel et al. [38] reported improved weed control with mixtures of mesotrione plus acetochlor or atrazine over that of mesotrione alone. As seen in this study, the combination of mesotrione with metolachlor plus atrazine has enhanced weed control in other studies [38].

| | | 2013 | | 2015 | |
|---|---|---|---|---|---|
| **Treatment** | **Dose** | **Taylor** | **Ganado** | **Taylor** | **Ganado** |
| | Kg ai ha$^{-1}$ | Days after treatment | | | |
| | | 95 | 109 | 101 | 44 |
| | | % | | | |
| Atrazine (A) | 1.1 | 73 | 99 | 79 | 72 |
| Fluthiacet-methyl (F)+pyroxasulfone (P) | 0.006 + 0.2 | - | - | 99 | 98 |
| (F)+(P)+(A) | 0.004 + 0.2 + 1.3 | - | - | 40 | 93 |
| Isoxaflutole | 0.05 | 100 | 98 | 51 | 67 |
| *S*-metolachlor (S) | 1.35 | 100 | 76 | 99 | 69 |
| (S)+(A) | 1.4 + 1.8 | 100 | 99 | 99 | 92 |
| Alachlor+(A) | 2.5 + 1.5 | 100 | 99 | - | - |
| Mesotrione (M) | 0.1 | 99 | 61 | 71 | 52 |
| Thiencarbazone-methyl+isoxaflutole | 0.02 + 0.06 | 100 | 92 | 99 | 83 |
| (S)+(A)+(M)+bicyclopyrone | 0.4 + 0.8 | - | - | 99 | 72 |
| | 0.09 + 0.02 | | | | |
| Acetochlor+(A) | 2.1 + 1.3 | 100 | 100 | 99 | 93 |
| Dimethenamid-P+(A) | 1.6 + 3.2 | 100 | 100 | 99 | 95 |
| Acetochlor (74.8%) | 6.9 | - | - | 100 | 100 |
| Rimsulfuron+(M) | 0.02 + 0.2 | 100 | 27 | - | - |
| Rimsulfuron+thifensulfuron-methyl | 0.02 + 0.02 | 99 | 90 | 98 | 37 |
| (S)+(A)+(M) | 1.5 + 1.5 + 0.2 | 100 | 97 | 99 | 90 |

| | | 2013 | | 2015 | |
|---|---|---|---|---|---|
| **Treatment** | **Dose** | **Taylor** | **Ganado** | **Taylor** | **Ganado** |
| Dimethenamid-P | 0.8 | 98 | 53 | 96 | 92 |
| Pendimethalin | 1.6 | 97 | 83 | 98 | 47 |
| Saflufenacil | 0.05 | 99 | 72 | 70 | 73 |
| Saflufenacil+dimethenamid-P | 0.08 + 0.7 | 100 | 95 | 99 | 100 |
| Acetochlor (33%) | 1.7 | 100 | 50 | 99 | 88 |
| (S)+(M) | 2.8 + 0.3 | 100 | 91 | 100 | 94 |
| (P) | 0.12 | 100 | 91 | 99 | 84 |
| Untreated | - | 0 | 0 | 0 | 0 |
| LSD (0.05) | | 17 | 27 | 22 | 24 |

**Table 3.** Palmer amaranth control in corn with PRE herbicides.

In 2013, thiencarbazone-methyl plus isoxaflutole provided perfect control (100%) of hophornbeam copperleaf while acetochlor (33%), saflufenacil or pyroxasulfone alone, alachlor plus atrazine, rimsulfuron plus thifensulfuron-methyl, *S*-metolachlor plus atrazine plus mesotrione, saflufenacil plus dimethenamid-P, and *S*-metolachlor plus mesotrione controlled this weed at least 92% (**Table 4**). Atrazine and mesotrione alone and rimsulfuron plus mesotrione provided unacceptable control (<60%). In 2014, either acetochlor (74.8%), isoxaflutole, saflufenacil, or pyroxasulfone alone controlled hophornbeam copperleaf at least 93% (**Table 4**). The combinations of *S*-metolachlor plus either atrazine or mesotrione and saflufenacil plus dimethenamid-P controlled this weed 90–98%, while rimsulfuron plus either mesotrione or thifensulfuron-methyl and acetochlor (33%) provided 67–70% control.

In 2013, under low common sunflower pressure (2–3 plants $m^2$), all herbicides, with the exception of atrazine alone (73%) and rimsulfuron plus thifensulfuron-methyl (87%), controlled this weed at least 95% (**Table 4**). In 2014, under slightly greater common sunflower populations (4–6 plants $m^2$), control was more variable. Acetochlor (74.8%) alone, thiencarbazone-methyl plus isoxaflutole, rimsulfuron plus thifensulfuron-methyl, saflufenacil plus dimethenamid-P, and *S*-metolachlor plus mesotrione controlled this weed at least 97%. Mesotrione, pendimethalin, or pyroxasulfone alone provided unacceptable control (<60%). The development of ALS-resistant common sunflower has limited the options for growers having to control common sunflower with POST herbicides [83, 84]. Results from this study are consistent with previous findings which found that common sunflower control with herbicide systems containing isoxaflutole was at least 85% in most instances [84, 85].

### *3.1.3. Corn injury and yield*

Grain yields were obtained only in 2013 at both locations and in 2015 at Taylor. Early season crop injury consisted of stunting and was never more than 3% with any herbicide treatment (data not shown). Corn recovered from the slight early season stunting and typically by harvest

| | | Hophornbeam copperleaf | | Common sunflower | |
|---|---|---|---|---|---|
| | | 2013 | 2014 | 2013 | 2014 |
| | | Days after treatment | | | |
| Treatment | Dose | 95 | 109 | 95 | 48 |
| | Kg ai ha$^{-1}$ | % | | | |
| Atrazine (A) | 1.1 | 38 | 80 | 73 | 77 |
| Isoxaflutole (I) | 0.05 | 77 | 98 | 100 | 79 |
| *S*-metolachlor (S) | 1.25 | 79 | 83 | 97 | 77 |
| (S)+(A) | 1.4 + 1.8 | 76 | 90 | 100 | 85 |
| Alachlor+(A) | 2.5 + 1.5 | 93 | - | 98 | - |
| Mesotrione (M) | 0.1 | 55 | 60 | 99 | 60 |
| Thiencarbazone-methyl+(I) | 0.02 + 0.06 | 100 | 77 | 98 | 100 |
| Acetochlor+(A) | 2.1 + 1.3 | 79 | - | 99 | - |
| Dimethenamid–P (D) | 0.8 | 72 | 82 | 97 | 79 |
| (D)+(A) | 1.7 + 3.2 | 80 | 72 | 97 | 93 |
| Acetochlor (74.8% formulation) | 6.8 | - | 99 | - | 97 |
| Rimsulfuron (R)+(M) | 0.02 + 0.2 | 60 | 67 | 97 | 93 |
| (R)+thifensulfuron-methyl | 0.02 + 0.02 | 98 | 70 | 87 | 99 |
| (S)+(A)+(M) | 1.5 + 1.5 + 0.2 | 98 | 74 | 100 | 87 |
| Pendimethalin | 1.6 | 69 | 85 | 95 | 58 |
| Saflufenacil (Sa) | 0.05 | 96 | 98 | 100 | 90 |
| (Sa)+(D) | 0.08 + 0.7 | 99 | 98 | 100 | 97 |
| Acetochlor (33% formulation) | 1.7 | 92 | 63 | 100 | 72 |
| (S)+(M) | 2.8 + 0.3 | 92 | 93 | 99 | 98 |
| Pyroxasulfone | 0.12 | 96 | 93 | 97 | 55 |
| Untreated | - | 0 | 0 | 0 | 0 |
| LSD (0.05) | | 34 | 30 | 20 | 36 |

**Table 4.** Hophornbeam copperleaf and common sunflower control in corn with PRE herbicides.

no differences in corn plant growth between the untreated check and any herbicide treatments were noted (data not shown). Although no appreciable crop injury was noted in these studies, this is not always true. Instances of isoxaflutole phytotoxicity in corn have been documented [85, 86] and attributed to several factors, including application timing [87], increased use dose [37], and varied susceptibility of corn hybrids to isoxaflutole [88]. Environmental factors (cool and wet) and soil characteristics [89] can also lead to corn injury by isoxaflutole. Johnson et al. [85] reported that PPI herbicide applications resulted in greater injury than PRE applications, and this was probably due to increased amount of precipitation. Armel et al. [38] reported that acetochlor, atrazine, or mesotrione combinations did cause 11–18% corn stunting when followed by 32 mm of rainfall, but that the corn recovered quickly and by 4 weeks after treatment injury did not exceed 2%.

In 2013 at the Taylor location, atrazine, isoxaflutole, and pyroxasulfone alone, *S*-metolachlor plus atrazine and/or mesotrione produced grain yields that were greater than the untreated check (**Table 5**). Although not significant, all herbicide treatments resulted in a numerical increase in grain yield over the untreated check. At the Ganado location, grain yields from the herbicide treatments were not significantly different from the untreated check; however, all yields from the herbicide treatments were numerically greater than the untreated check with the exception of *S*-metolachlor plus mesotrione which produced a 10% decrease in yield from the untreated check. No reason for this reduction can be determined.

In 2014, no significant differences between the untreated check and any herbicide treatments were noted, although several herbicide treatments produced numerically greater yields than the untreated check (**Table 5**). Dimethenamid-P and pyroxasulfone alone, fluthiacet-methyl plus pyroxasulfone, thiencarbazone-methyl plus isoxaflutole, dimethenamid-P plus atrazine, *S*-metolachlor plus atrazine plus mesotrione, and saflufenacil plus dimethenamid-P produced grain yields that were 14–21% greater than the untreated check.

| | | 2013 | | 2015 |
|---|---|---|---|---|
| **Herbicide treatment** | **Dose** | **Taylor** | **Ganado** | **Taylor** |
| | **Kg ai ha$^{-1}$** | **Kg ha$^{-1}$** | | |
| Atrazine (A) | 1.1 | 5586 | 7695 | 7556 |
| Fluthiacet-methyl (FM)+pyroxasulfone (P) | 0.006+0.2 | - | - | 9342 |
| (F)+(P)+(A) | 0.004+0.2+1.3 | - | - | 8092 |
| *S*-metolachlor (S) | 1.3 | 5143 | 7082 | 8806 |
| Isoxaflutole (I) | 0.05 | 5434 | 6980 | 7669 |
| (S)+(A) | 1.4+1.8 | 5396 | 7627 | 8582 |
| Alachlor+(A) | 2.5+1.5 | 4940 | 7466 | - |
| Mesotrione (M) | 0.1 | 4851 | 7727 | 8970 |
| Thiencarbazone-methyl+(I) | 0.02+0.06 | 5256 | 7318 | 9494 |
| Acetochlor+(A) | 2.1+1.3 | 4915 | 7031 | 8899 |
| Dimethenamid-P (D) | 0.8 | 5275 | 7172 | 9447 |
| (D)+(A) | 1.6+3.2 | 5294 | 8350 | 9611 |
| Acetochlor (74.8%) | 6.9 | - | - | 8738 |
| Rimsulfuron (R)+(M) | 0.02+0.2 | 4972 | 8295 | - |
| (R)+thifensulfuron-methyl | 0.02+0.02 | 5168 | 7991 | 7934 |

| | | 2013 | | 2015 |
|---|---|---|---|---|
| Herbicide treatment | Dose | Taylor | Ganado | Taylor |
| | Kg ai $ha^{-1}$ | Kg $ha^{-1}$ | | |
| (S)+(A)+(M) | 1.5+1.5+0.2 | 5589 | 8556 | 9962 |
| Pendimethalin | 1.6 | 5264 | 7881 | 8958 |
| Saflufenacil | 0.05 | 4524 | 8311 | 7477 |
| Saflufenacil+dimethenamid-P | 0.08+0.7 | 4906 | 7495 | 9377 |
| Acetochlor (33%) | 1.7 | 5099 | 8310 | 8691 |
| (S)+(M) | 2.8+0.3 | 5501 | 6160 | 8695 |
| Pyroxasulfone | 0.1 | 5346 | 7548 | 9691 |
| Untreated | - | 4506 | 6816 | 8218 |
| LSD (0.05) | | 796 | 1800 | 1969 |

**Table 5.** Corn yield as influenced by PRE herbicides.

With glyphosate-resistant pigweed becoming more widespread throughout the state, the use of soil-applied herbicides can not only control resistant weed species in glyphosate-resistant corn production systems but can also reduce the risk of new herbicide-resistant weed species occurring. In general, many treatments with two or three herbicide modes of action provided better weed control than one herbicide alone, and the chance of corn injury appears to be minimal with any herbicide combinations under normal growing conditions. Our results indicate that in a year with little or no rainfall within 7–14 days after PRE herbicide application, any combination of PRE herbicides may need to be followed by POST herbicides for control of escaped weeds.

### 3.2. Corn POST Studies

#### 3.2.1. Annual grass control

Limited control of browntop panicum was noted when using POST herbicides. Glyphosate and tembotrione alone provided 99% browntop panicum control, while the combinations of atrazine plus *S*-metolachlor plus glyphosate, mesotrione plus *S*-metolachlor plus glyphosate, and thiencarbazone-methyl plus tembotrione provided 96–98% control (**Table 6**). Mesotrione and topramezone alone and the combination of primisulfuron-methyl plus pyroxasulfone controlled this weed 77–83%; however, no other herbicides provide better than 68% control. Stephenson et al. [13] noted that thiencarbazone plus tembotrione controlled browntop millet (*Urochloa ramose* L.) 93% which was greater than tembotrione, atrazine, or glufosinate alone. They also noted that the co-application of atrazine, glufosinate, or glyphosate with thiencarbazone plus tembotrione did not increase browntop millet control.

In 2014 at Beasley only nicosulfuron, primisulfuron-methyl, and topramezone alone or the combination of pyroxasulfone plus glyphosate provided acceptable Texas millet control (>84%), while at Beyersville only the combinations of mesotrione plus *S*-metolachlor plus glyphosate and fluthiacet-methyl plus pyroxasulfone plus atrazine controlled this weed at least 81% (**Table 6**). Prostko et al. [88] found that glyphosate applied sequentially was more effective at controlling Texas millet than either nicosulfuron or foramsulfuron. Again, the added control noted with pyroxasulfone can be attributed to the extended residual activity of this herbicide [81].

The combinations of atrazine plus *S*-metolachlor plus mesotrione plus bicyclopyrone, atrazine plus *S*-metolachlor plus glyphosate, dimethenamid plus glyphosate, fluthiacet-methyl plus pyroxasulfone plus glyphosate, mesotrione plus *S*-metolachlor plus glyphosate, pyroxasulfone plus glyphosate, and thiencarbazone-methyl plus tembotrione controlled barnyardgrass at least 93% (**Table 6**). Lamore et al. [89] reported that tembotrione at 92 g $ha^{-1}$ provided greater than 90% control, which is similar to the results in this study. Stephenson et al. [13] reported that thiencarbazone plus tembotrione or tembotrione alone provided equivalent control of barnyardgrass to atrazine plus either glufosinate or glyphosate.

| | | **2013** | **2014** | | **2015** |
|---|---|---|---|---|---|
| | **Dose** | **Browntop panicum** | **Texas millet[g]** | | **Barnyardgrass** |
| **Herbicide treatment** | **Kg ai or ae $ha^{-1}$** | **Taylor** | **Bea** | **Beyers[h]** | **Taylor** |
| | | Days after treatment | | | |
| | | 98 | 69 | 93 | 53 |
| | | % | | | |
| Atrazine (A)[c] | 1.1 | 45 | - | 58 | 70 |
| Carfentrazone-ethyl[a, d] | 0.02 | 66 | - | - | 60 |
| Fluroxypyr[a, c] | 0.3 | 64 | - | 68 | 40 |
| Fluthiacet-methyl (FM)[a, c] | 0.07 | 43 | - | 53 | 39 |
| Glufosinate ammonium[a] | 0.7 | - | - | - | 85 |
| Glyphosate (G) | 1.5 ae | 99 | 25 | 63 | 85 |
| Halosulfuron-methyl (HM) [a, c] | 0.07 | 49 | - | 72 | 20 |
| Mesotrione (M)[a, c] | 0.1 | 79 | - | 57 | 58 |
| Nicosulfuron[b, d] | 0.04 | - | 84 | - | - |
| Primisulfuron-methyl (PM)[b, d] | 0.04 | 68 | 89 | - | - |
| Prosulfuron[c] | 0.04 | 65 | - | 40 | 20 |
| Tembotrione[a, e] | 0.09 | 99 | 37 | 73 | 87 |
| Topramezone[a, c] | 0.15 | 77 | 88 | 77 | - |
| (A)+*S*-metolachlor (S)+(M)+ bicyclopyrone[a, c] | 0.7+1.5+ 0.17+0.04 | - | - | - | 93 |
| (A)+(S)+(G)[a, c] | 1.8+1.5+0.8 ae | 98 | 56 | 73 | 99 |
| Diflufenzopyr+dicamba (D)[a, c] | 0.06+0.1 ae | 59 | - | - | 50 |
| Dimethenamid+(G) | 0.8+1.54 ae | - | - | 73 | 99 |

| | | 2013 | 2014 | | 2015 |
|---|---|---|---|---|---|
| | Dose | Browntop panicum | Texas millet[g] | | Barnyardgrass |
| Herbicide treatment | Kg ai or ae $ha^{-1}$ | Taylor | Bea | Beyers[h] | Taylor |
| (FM)+(M)[a, c] | 0.09+0.09 | - | - | - | 55 |
| (FM)+pyroxasulfone (P)+(A)[d] | 0.004+0.2+1.3 | - | - | 96 | 62 |
| (FM)+(P)+(G)[d] | 0.004+0.2+1.5 ae | - | - | - | 100 |
| (HM)+(D)[a, c] | 0.07+0.1 ae | 52 | - | 72 | 40 |
| (M)+(S)+(G)[a, c] | 0.1+1.1+1.1 ae | 96 | 55 | 81 | 98 |
| (PM)+(P)[a, c] | 0.03 + 0.01 | 83 | 74 | 70 | 43 |
| Pyroxasulfone+(G) | 0.1+1.5 ae | - | 96 | 53 | 96 |
| Thiencarbazone-methyl+tembotrione[a, c] | 0.02+0.07 | 98 | 53 | 53 | 98 |
| Untreated | - | 0 | 0 | 0 | 0 |
| LSD (0.05) | | 34 | 18 | 20 | 28 |

[a] AMS (ammonium sulfate) at 3.86 kg/378.4 L.
[b] UAN (urea-ammonium nitrate) added at 2.2 L.
[c] Crop oil concentrate (Agridex) added at 1.0% v/v.
[d] Non-ionic surfactant (Induce) added at 0.25% v/v.
[e] Methylated seed oil (Phase) added at 1.1 L.
[f] Grass height at application: Taylor, ≤ 15 cm; Besley, ≤ 10 cm; Coupland, ≤ 5 cm; Taylor, ≤ 15 cm.
[g] Texas millet locations: Bea, Beasley; Beyers, Beyersville.
[h] Glyphosate at 1.54 kg ae $ha^{-1}$ added to all treatments with the exception of glufosinate ammonium and glyphosate alone.

**Table 6.** Annual grass control in corn with POST herbicides[f].

### *3.2.2. Broadleaf weed control*

In 2014 at Yoakum, under dense Palmer amaranth populations, atrazine, prosulfuron, and topramezone alone or the combinations of atrazine plus *S*-metolachlor plus glyphosate, diflufenzopyr plus dicamba, dimethenamid plus glyphosate, halosulfuron-methyl plus dicamba, mesotrione plus *S*-metolachlor plus glyphosate, pyroxasulfone plus glyphosate, and thiencarbazone-methyl plus tembotrione provided at least 91% control (**Table 7**). Armel et al. [38] reported improved weed control with mixtures of mesotrione plus acetochlor or atrazine.

At the Taylor location in 2015, under low populations, only carfentrazone-ethyl, fluroxypyr, fluthiacet-methyl, and primisulfuron-methyl plus pyroxasulfone failed to provide at least 85% Palmer amaranth control. At the Ganado location, only pyroxasulfone plus glyphosate controlled this weed at least 80%, and this general lack of control was probably due to weed height (40–60 cm) at the time of herbicide application. Herbicide application to weeds 10–15 cm tall can result in corn grain yields equal to those in weed-free plots [90], but POST applications when weeds are greater than 15 cm tall provided inconsistent season-long weed control when compared with applications when weeds are less than 15 cm tall [91]. Stephenson et al. [13] reported that atrazine alone provided 96% control of this weed, while thiencarbazone plus tembotrione or tembotrione, glufosinate, and glyphosate alone provided 92% or less control.

Glyphosate alone provided 100% pitted morningglory control, while mesotrione plus *S*- meto-lachlor plus glyphosate controlled this weed 82% (**Table 8**). Typically, glyphosate provides inadequate control of pitted morningglory when applied alone at normal label use doses [92–94]. However, greater than 90% late season control of tall morningglory (*Ipomoea purpurea* L.), ivyleaf morningglory (*I. hederacea* L.), and entireleaf morningglory (*I. hederacea* var. *integriuscula* Gray) in the field has been documented with 1.12 kg ha$^{-1}$ of glyphosate applied to plants with six true leaves or less [95]. However, sequential in-season glyphosate applications are often required to provide similar levels of pitted morningglory control [96, 97]. No other herbicides provided better than 68% control. Bararpour et al. [98] observed 90–100% control of entireleaf and pitted morningglory with the combination of thiencarbazone plus tembotrione plus either atrazine, glufosinate, or glyphosate, while Stephenson et al. [13] observed 85–88% control with thiencarbazone plus tembotrione alone.

| | **Dose** | **2014** | **2015** | |
|---|---|---|---|---|
| **Herbicide treatment** | **Kg ai or ae ha$^{-1}$** | **Yoakum** | **Taylor** | **Ganado[g]** |
| | | Days after treatment | | |
| | | 40 | 53 | 32 |
| | | % | | |
| Atrazine (A)[c] | 1.1 | 100 | 100 | 43 |
| Carfentrazone-ethyl[a, d] | 0.02 | 67 | 20 | 25 |
| Fluroxypyr[a, c] | 0.3 | 48 | 70 | 10 |
| Fluthiacet-methyl (FM)[a, c] | 0.07 | 58 | 54 | 33 |
| Glufosinate ammonium[a] | 0.7 | 63 | 100 | 58 |
| Glyphosate (G) | 1.5 ae | 68 | 100 | 59 |
| Halosulfuron-methyl (HM)[a, c] | 0.07 | 53 | 85 | 13 |
| Mesotrione (M)[a, c] | 0.1 | 83 | 100 | 55 |
| Prosulfuron[c] | 0.04 | 91 | 98 | 33 |
| Tembotrione[a, e] | 0.1 | 83 | 100 | 63 |
| Topremazone[a, c] | 0.15 | 97 | 100 | 63 |
| (A)+*S*-metolachlor(S)+(M) +bicyclopyrone[a, c] | 0.8+1.5+0.2+0.04 | - | 99 | 75 |
| (A)+(S)+(G)[a, c] | 1.8+1.5+0.8 ae | 100 | 99 | 73 |
| Diflufenzopyr+dicamba (D)[a, c] | 0.06+0.14 ae | 91 | 99 | 65 |
| Dimethenamid-P+(G) | 0.8+1.5 ae | 100 | 100 | 67 |
| (FM)+(M)[a, c] | 0.09+0.09 | - | 100 | 47 |

| | Dose | 2014 | 2015 | |
|---|---|---|---|---|
| Herbicide treatment | Kg ai or ae ha$^{-1}$ | Yoakum | Taylor | Ganado[g] |
| (FM)+pyroxasulfone (P)+(G)[d] | 0.004+0.15+1.3 ae | - | 100 | 69 |
| (FM)+(P)+(A)[d] | 0.004+0.15+1.5 | - | 99 | 43 |
| (HM)+(D)[a, c] | 0.07+0.3 ae | 99 | 100 | 60 |
| (M)+(S)+(G)[a, c] | 0.1+1.1+1.1 ae | 100 | 100 | 60 |
| Primisulfuron-methyl+(P)[b, c] | 0.03+0.01 | 67 | 78 | 7 |
| Pyroxasulfone+(G) | 0.1+1.5 ae | 99 | 100 | 80 |
| Thiencarbazone-methyl+tembotrione[a, c] | 0.02 + 0.07 | 100 | 100 | 58 |
| Untreated | - | 0 | 0 | 0 |
| LSD (0.05) | | 16 | 22 | 25 |

[a] AMS (ammonium sulfate) added at 3.86 kg/378.4 L.
[b] UAN (urea-ammonium nitrate) added at 2.2 L.
[c] Crop oil concentrate (Agridex) added at 1.0% v/v.
[d] Non-ionic surfactant (Induce) added at 0.25% v/v.
[e] Methylated seed oil (Phase) added at 1.1 L.
[f] *A. palmeri* height at application: Yoakum, ≤ 7.6 cm; Taylor, ≤ 10 cm; Ganado, ≤ 61 cm.
[g] Glyphosate at 1.54 kg ae ha$^{-1}$ added to all treatments with the exception of glyphosate and glufosinate ammonium alone.

**Table 7.** Palmer amaranth control in corn with POST herbicides[f].

| | | 2013 | | 2015 |
|---|---|---|---|---|
| Herbicide treatment | Dose | Pitted morningglory | Hophornbeam copperleaf | Asiatic dayflower |
| | Kg ai or ae ha$^{-1}$ | Days after treatment | | |
| | | 60 | 60 | 31 |
| | | % | | |
| Atrazine (A)[c] | 1.1 | 12 | 85 | 45 |
| Carfentrazone-ethyl[a, d] | 0.02 | 15 | 84 | - |
| Dicamba (D)[c] | 0.56 | 20 | 54 | - |
| Fluroxypyr[a, c] | 0.3 | 20 | 40 | - |
| Fluthiacet-methyl (FM)[a, c] | 0.07 | 7 | 91 | 55 |
| Glufosinate ammonium[a] | 0.7 | - | - | 90 |
| Glyphosate (G) | 1.5 ae | 100 | 79 | 62 |
| Halosulfuron-methyl (HM) [a, c] | 0.07 | 13 | 73 | - |
| Mesotrione (M)[a, c] | 0.1 | 47 | 64 | 84 |

| | | 2013 | | 2015 |
|---|---|---|---|---|
| **Herbicide treatment** | **Dose** | **Pitted morningglory** | **Hophornbeam copperleaf** | **Asiatic dayflower** |
| | **Kg ai or ae ha$^{-1}$** | **Days after treatment** | | |
| Primisulfuron-methyl (PM)[b, d] | 0.04 | 50 | 65 | - |
| Prosulfuron[c] | 0.04 | 25 | 86 | - |
| Tembotrione[a, e] | 0.1 | 53 | 90 | 65 |
| Topramezone[a, c] | 0.15 | 36 | 74 | 82 |
| A + *S*-metolachlor (S)+(M)+ bicyclopyrone[a, c] | 0.8+1.5+0.2+0.04 | - | - | 93 |
| (A)+(S)+(G)[a, c] | 1.8+1.5+0.8 ae | 63 | 99 | 93 |
| Diflufenzopyr+(D)[a, c] | 0.06+0.14 ae | 30 | 62 | 79 |
| (FM)+(M)[a, c] | 0.09+0.09 | - | - | 78 |
| (FM)+pyroxasulfone (P)+ (A)[d] | 0.004+0.15+1.3 | - | - | 70 |
| (HM)+(D)[a, c] | 0.07+0.3 ae | 20 | 40 | 83 |
| (M)+(S)+(G)[a, c] | 0.1+1.1+1.1 ae | 82 | 92 | 81 |
| Primisulfuron-methyl+(P)[b, c] | 0.03+0.01 | 3 | 76 | - |
| Thiencarbazone-methyl+tembotrione[a, c] | 0.02+0.07 | 68 | 81 | 72 |
| Untreated | - | 0 | 0 | 0 |
| LSD (0.05) | | 30 | 25 | 18 |

[a] AMS (ammonium sulfate) added at 3.86 kg/378.4 L.
[b] UAN (urea-ammonium nitrate) added at 2.2 L.
[c] Crop oil concentrate (Agridex) added at 1.0% v/v.
[d] Non-ionic surfactant (Induce) added at 0.25% v/v.
[e] Methylated seed oil (Phase) added at 1.1 L.
[f] Pitted morningglory height at application, ≤ 20 cm; hophornbeam copperleaf, ≤ 15 cm;Asiatic dayflower ≤ 7.6 cm.

**Table 8.** Broadleaf weed control in corn with POST herbicides[f].

Fluthiacet-methyl and tembotrione alone and the combinations of atrazine plus *S*-metolachlor plus glyphosate and mesotrione plus *S*-metolachlor plus glyphosate provided at least 90% hophornbeam copperleaf control, while atrazine, carfentrazone-ethyl, and prosulfuron alone and the combination of thiencarbazone-methyl plus tembotrione provided 81–86% control (**Table 8**).

Glufosinate ammonium alone controlled Asiatic dayflower 90%, while the combinations of atrazine plus *S*-metolachlor plus mesotrione plus bicyclopyrone and atrazine plus *S*-metolachlor plus glyphosate provided 93% control (**Table 8**).

### *3.2.3. Corn injury and yield*

Crop injury consisted of stunting with some leaf chlorosis and necrosis and was never more than 8% with any herbicide treatment (data not shown). Corn recovered from the slight early season stunting and typically by harvest no differences in corn plant growth between the untreated check and any herbicide treatments were noted (data not shown). Although no appreciable crop injury was noted in these studies this is not always true. Other studies have reported corn injury more than 50% with isoxaflutole, imazethapyr, imazapic, and prosulfuron in field or sweet corn [99–101]. In addition, herbicides such as halosulfuron and dicamba plus diflufenzopyr have been reported to cause as much as 25 and 15% injury, respectively [102, 103]. Corn phytotoxicity has been attributed to several factors, including application timing [104], increased use doses [93], and varied susceptibility of corn hybrids to different herbicides [105].

Corn yield was combined over locations due to a lack of treatment by location interaction. Yields were likely affected more by weed control than any other factor (rainfall, etc.) in any year. Pyroxasulfone plus glyphosate produced the greatest yield while halosulfuron alone and the untreated check produced the least yield (**Table 9**). Treatments that contained the combination of atrazine plus glyphosate resulted in yields that were greater than 5200 kg ha$^{-1}$.

Some research suggests that timely POST control can be an effective alternative to soil-applied herbicides in corn [45, 105, 106]. The use of POST herbicides only is generally considered a greater risk and requires careful management [45, 105, 106]. Also, weed density and application timing are factors in weed efficacy with POST herbicides. Halford et al. [107] reported a reduction in yield when weeds remained beyond V6 corn. In addition, Gower et al. [106] found that subsequent emergence and competition after early glyphosate applications was likely responsible for corn yield reductions. Also, late POST applications can reduce corn grain yields, although weed control was nearly perfect [105, 108].

| **Herbicide treatment** | **Dose** | **Yield** |
|---|---|---|
| | **Kg ai or ae ha$^{-1}$** | **Kg ha$^{-1}$** |
| Atrazine (A)[c] | 1.1 | 2887 |
| Carfentrazone-ethyl[a, d] | 0.02 | 2253 |
| Fluroxypyr[a, c] | 0.3 | 1375 |
| Fluthiacet-methyl (FM)[a, c] | 0.07 | 3270 |
| Glufosinate ammonium[a] | 0.7 | 2724 |
| Glyphosate (G) | 1.5 ae | 4413 |
| Halosulfuron-methyl (HM) [a, c] | 0.07 | 942 |
| Mesotrione (M)[a, c] | 0.1 | 5568 |
| Prosulfuron[c] | 0.04 | 2266 |
| Tembotrione[a, e] | 0.1 | 4199 |
| Topramezone[a, c] | 0.15 | 3164 |
| (A) + S-metolachlor + (M) + bicyclopyrone[a, c] | 0.8+1.5+0.2+ 0.04 | 5248 |
| (A) + (S) + (G)[a, c] | 1.8+1.5+ 0.8 ae | 5587 |
| Diflufenzopyr + (D)[a, c] | 0.06+0.14 | 1601 |
| Dimethenamid-P + (G) | 0.8+1.5 ae | 4425 |

| Herbicide treatment | Dose | Yield |
|---|---|---|
| | Kg ai or ae ha$^{-1}$ | Kg ha$^{-1}$ |
| (FM) + (M)[a, c] | 0.09+0.09 | 4909 |
| (FM) + pyroxasulfone (P) + (A)[d] | 0.004+0.15+1.3 | 2178 |
| (HM) + (D)[a, c] | 0.07+0.3 ae | 3490 |
| (M) + (S) + (G)[a, c] | 0.1+1.1+1.1 ae | 4149 |
| (PM) + (P)[a, c] | 0.03 + 0.01 | 1506 |
| (P) + (G) | 0.1+1.5 ae | 5781 |
| Thiencarbazone-methyl + tembotrione[a, c] | 0.02+0.07 | 4281 |
| Untreated | - | 395 |
| LSD (0.05) | | 2988 |

[a] AMS (ammonium sulfate) added at 3.86 kg/378.4 L.
[b] UAN (urea-ammonium nitrate) added at 2.2 L.
[c] Crop oil concentrate (Agridex) added at 1.0% v/v.
[d] Non-ionic surfactant (Induce) added at 0.25% v/v.
[e] Methylated seed oil (Phase) added at 1.1 L.

**Table 9.** Corn yield as influenced by POST herbicides.

With glyphosate-resistant pigweed becoming more widespread throughout the state, the use of POST herbicide combinations, which may or may not contain glyphosate, can not only control resistant weed species in glyphosate-resistant corn production systems but can also reduce the risk of new herbicide-resistant weed species occurring. In general, many treatments with two or three herbicides with different modes of action provided better weed control than one herbicide alone, and the chance of corn injury appears to be minimal with any herbicide combination under normal growing conditions.

## 3.3. Cotton studies

### *3.3.1. South-central Texas*

A significant year-by-treatment interaction existed for all weed control and cotton yield data, thus data were analyzed separately by year. Weed control data required arcsine transformation in order to meet the assumption of homogeneity of variances for ANOVA; however, the non-transformed means are reported in the (**Table 10–13**).

In 2012, control of Palmer amaranth ranged from 29 to 97% while common waterhemp control ranged from 55 to 100% prior to EPOST applications (**Table 10**). At that timing, control of Palmer amaranth was lowest with pyrithiobac applied PRE. Similar results were seen for common waterhemp control, where pyrithiobac applied PRE provided only 55% control. After EPOST and MPOST applications, no differences in Palmer amaranth control were detected among treatments, with means ranging from 93 to 100%. After the EPOST application timing, control of common waterhemp with *S*-metolachlor plus fomesafen applied PRE was lower than control provided by trifluralin applied PPI followed by either glyphosate plus dicamba- or glufosinate plus dicamba applied EPOST, and all treatments that included glyphosate plus dicamba plus acetochlor applied EPOST. After MPOST applications, no differences in common waterhemp control among herbicide treatments were observed. Seed cotton yields of treated plots ranged from 3581 to 4002 kg ha$^{-1}$, which were all greater than the non-treated check (1823 kg ha$^{-1}$).

In 2013, Palmer amaranth control prior to EPOST application ranged from 63 to 100%, while control of common waterhemp ranged from 60 to 100% (**Table 11**). Similar to 2012, pyrithiobac applied PRE resulted in the lowest control of both Palmer amaranth and common waterhemp (63 and 60%, respectively). Treatments that included pendimethalin-applied PRE provided reduced control of Palmer amaranth (88–90%) when compared with many other treatments at the early rating. A similar pattern was observed with common waterhemp, where control was numerically lower from treatments of pendimethalin-applied PRE than many other treatments,

| **Herbicide and application timing**[a] | | | | **Palmer amaranth** | | | **Common waterhemp** | | | **Seed cotton** |
|---|---|---|---|---|---|---|---|---|---|---|
| **PPI** | **PRE** | **EPOST** | **MPOST** | **Early** | **Mid** | **Late** | **Early** | **Mid** | **Late** | **Yield** |
| | | | | % | | | | | | **Kg ha$^{-1}$** |
| | P | [A] | | 83 | 99 | 99 | 73 | 99 | 98 | 3779 |
| | P | [A]+D+Ace | | 83 | 100 | 100 | 95 | 100 | 100 | 3966 |
| | P | Pyr | Gluf | 84 | 95 | 99 | 79 | 89 | 100 | 3647 |
| | Pyr | [A]+D+Ace | | 29 | 99 | 100 | 55 | 100 | 100 | 3728 |
| | *S* | | [B]+Trif | 96 | 99 | 100 | 95 | 97 | 100 | 3955 |
| | F | | [B]+Trif | 88 | 93 | 100 | 92 | 89 | 100 | 3721 |
| | *S*+F | | [B]+Trif | 85 | 99 | 100 | 91 | 85 | 100 | 3680 |
| | *S*+Pr | | [B]+Trif | 81 | 99 | 100 | 89 | 91 | 100 | 3673 |
| T | | Gluf | Gluf | 86 | 100 | 100 | 98 | 99 | 100 | 3581 |
| T | | [A]+D | [A] | 97 | 100 | 100 | 100 | 100 | 100 | 3859 |
| T | | Gluf+D | [A]+D | 99 | 100 | 100 | 100 | 100 | 100 | 3779 |
| T | | [A]+D+Ace | | 96 | 100 | 100 | 100 | 100 | 100 | 4002 |
| - | - | - | None | 0 | 0 | 0 | 0 | 0 | 0 | 1823 |
| LSD (0.05) | | | | 15 | 7 | 1 | 13 | 11 | 1 | 559 |

[a]Herbicide abbreviations, product name and doses: acetochlor, Warrant (Ace) at 1.26 kg ai ha$^{-1}$; dicamba, Clarity (D) at 0.56 kg ae ha$^{-1}$; glufosinate, Liberty (Gluf) at 0.59 kg ai ha$^{-1}$; fomesafen, Reflex (F) at 0.28 kg ai ha$^{-1}$;glyphosate (A), Roundup PowerMAX (Glyp [A]) at 1.26 kg ae ha$^{-1}$; glyphosate (B), Touchdown Total (Glyp [B]) at 0.88 kg ae ha$^{-1}$; pendimethalin, Prowl $H_20$ (P) at 1.6 kg ai ha$^{-1}$; prometryn, Caparol (Pr) at 0.56 kg ai ha$^{-1}$; pyrithiobac, Staple LX (Pyr) at 58.84 g ai ha$^{-1}$ PRE, 72.86 g ai ha$^{-1}$ POST; *S*-metolachlor, Dual Magnum (*S*) at 1.07 kg ai ha$^{-1}$; trifloxysulfuron, Envoke (Trif) at 5.25 g ai ha$^{-1}$; and trifluralin, Treflan (T) at 1.12 kg ai ha$^{-1}$.

**Table 10.** Palmer amaranth and common waterhemp control and seed cotton yield in 2012.

| Herbicide and application timing[a] | | | | Palmer amaranth | | | Common waterhemp | | | Seed cotton |
|---|---|---|---|---|---|---|---|---|---|---|
| PPI | PRE | EPOST | MPOST | Early | Mid | Late | Early | Mid | Late | Yield |
| | | | | % | | | | | | Kg ha$^{-1}$ |
| | P | [A] | | 88 | 100 | 100 | 88 | 100 | 99 | 4084 |
| | P | [A]+D+Ace | | 88 | 99 | 100 | 79 | 100 | 100 | 3838 |
| | P | Pyr | Gluf | 90 | 86 | 92 | 92 | 86 | 93 | 3526 |
| | Pyr | [A] + D+Ace | | 63 | 99 | 100 | 60 | 99 | 100 | 4034 |
| | *S* | | [B]+Trif | 93 | 82 | 99 | 91 | 81 | 98 | 3773 |
| | F | | [B]+Trif | 100 | 98 | 99 | 100 | 100 | 100 | 3986 |
| | *S*+F | | [B]+Trif | 100 | 100 | 100 | 100 | 100 | 100 | 4003 |
| | *S*+Pr | | [B]+Trif | 99 | 93 | 100 | 97 | 96 | 99 | 3881 |
| T | | Gluf | Gluf | 99 | 99 | 100 | 99 | 100 | 100 | 3983 |
| T | | [A]+D | [A] | 99 | 99 | 99 | 99 | 100 | 100 | 4122 |
| T | | Gluf + D | [A]+D | 99 | 100 | 100 | 98 | 100 | 100 | 4207 |
| T | | [A]+D+Ace | | 98 | 100 | 100 | 96 | 100 | 100 | 4209 |
| - | - | - | - | 0 | 0 | 0 | 0 | 0 | 0 | 254 |
| LSD (0.05) | | | | 8 | 4 | 2 | 8 | 6 | 5 | 530 |

[a]Herbicide abbreviations, product name and doses: acetochlor, Warrant (Ace) at 1.26 kg ai ha$^{-1}$; dicamba, Clarity (D) at 0.56 kg ae ha$^{-1}$; fomesafen, Reflex (F) at 0.28 kg ai ha$^{-1}$; glufosinate, Liberty (Gluf) at 0.59 kg ai ha$^{-1}$; glyphosate (A), Roundup PowerMAX (Glyp [A]) at 1.26 kg ae ha$^{-1}$; glyphosate (B), Touchdown Total (Glyp [B]) at 0.88 kg ae ha$^{-1}$; pendimethalin, Prowl $H_20$ (P) at 1.6 kg ai ha$^{-1}$; prometryn, Caparol (Pr) at 0.56 kg ai ha$^{-1}$; pyrithiobac, Staple LX (Pyr) at 58.84 g ai ha$^{-1}$ PRE, 72.86 g ai ha$^{-1}$ POST; S-metolachlor, Dual Magnum (*S*) at 1.07 kg ai ha$^{-1}$; trifloxysulfuron, Envoke (Trif) at 5.25 g ai ha$^{-1}$; and trifluralin, Treflan (T) at 1.12 kg ai ha$^{-1}$.

**Table 11.** Palmer amaranth and common waterhemp control and seed cotton yield in 2013.

though this was not always significant. Prior to the MPOST application, control of Palmer amaranth with *S*-metolachlor applied PRE was less than that of all other treatments except for pendimethalin applied PRE followed by pyrithiobac applied EPOST, which itself was lower than treatments other than *S*-metolachlor plus prometryn applied PRE. Control of common waterhemp prior to the MPOST application was lowest with *S*-metolachlor applied PRE (81%) and pendimethalin applied PRE followed by pyrithiobac applied EPOST (86%). At the last rating, control of Palmer amaranth and common waterhemp was reduced with pendimethalin applied PRE followed by pyrithiobac applied EPOST followed by glufosinate applied MPOST (92 and 93%, respectively) when compared to all other herbicide treatments. Mean yield of the non-treated control was 254 kg ha$^{-1}$, which was lower than that of all herbicide treatments (3526–4209 kg ha$^{-1}$).

Pyrithiobac applied PRE has been shown to provide satisfactory control of Amaranthus weeds [109, 110]; however, the opposite was observed in this experiment, where pyrithiobac applied PRE resulted in decreased levels of control of both Palmer amaranth and common waterhemp. The reasons for this lack of control are unknown, as the treatment was applied at the recommended rate and timing [111]. Pendimethalin applied PRE provided varied levels of control of common waterhemp, particularly in 2012. This may be due to the utilization of furrow irrigation for herbicide incorporation rather than overhead irrigation, which is recommended on the product label [112]. Trifluralin applied PPI consistently provided the best levels of control of both species. This is likely due in large part to the thorough mechanical incorporation of herbicide into the soil, which has been observed to affect the efficacy of trifluralin [113, 114].

In 2013, a decreased level of control of both Palmer amaranth and common waterhemp later in the season was observed with pendimethalin applied PRE followed by pyrithiobac-applied EPOST followed by glufosinate applied MPOST. This is attributed to a failure of glufosinate to control weeds that survived pyrithiobac EPOST and grew to a size larger than that recommended for control with glufosinate [115]. This weed size effect on glufosinate performance was observed by Craigmyle et al. [116], where control of common waterhemp was found to decrease with increasing plant height. Pendimethalin applied PRE followed by glyphosate EPOST provided excellent control of both species; however, in the presence of a glyphosate-resistant population, this treatment would likely not provide acceptable levels of control. In addition, this reliance on glyphosate as the single POST herbicide mechanism of action is not recommended due to the potential for selection of glyphosate-resistant plants [117]. Excellent control of both Palmer amaranth and common waterhemp was achieved in both years in treatments that included glyphosate plus dicamba plus acetochlor-applied EPOST. In addition to providing successful levels of weed control, this tank-mix would likely be very resilient against selecting resistant biotypes as suggested by Evans et al. [118], who found that the presence of glyphosate-resistant biotypes of common waterhemp was much less common in fields that received applications of mixed herbicide mechanisms of action.

#### *3.3.2. High Plains studies*

In the glyphosate plus 2,4-D choline study, trifluralin alone failed to control Palmer amaranth with only 20% control early season and no control late season, while systems which include POST applications of either glyphosate or glufosinate alone controlled this weed 23–53% (**Table 12**). Systems which included an EPOST and MPOST application of glyphosate plus 2,4-D choline provided at least 94% season-long control of this weed. Chahal and Johnson [119] reported that the addition of 2,4-D to glyphosate provided 99% control of glyphosate-resistant horseweed [*Conyza canadensis* (L.) Cronq.] compared to only 12% with glyphosate alone. In a similar study, 2,4-D added to glufosinate provided an increased level of common waterhemp control compared to herbicide treatments consisting of glufosinate only [116]. Miller and Norsworthy [120] reported that the addition of a residual herbicide, such as trifluralin, would provide an additional effective herbicide mode of action for managing resistant Palmer amaranth. Applications of 2,4-D, glyphosate, and glufosinate alone or tank-mixed represent broad-spectrum POST herbicides that have the potential to control 9 of the 10 most problematic weeds in the southern cotton and soybean production [121].

| Herbicide and application[a] | | | Palmer amaranth control | | Cotton injury | Lint yield |
|---|---|---|---|---|---|---|
| PPI[a] | EPOST | MPOST | Early | Late | Early | |
| | | | % | | | Kg ha$^{-1}$ |
| Trif | None | None | 20 | 0 | 0 | 0 |
| Trif | Gly | Gly | 46 | 23 | 0 | 0 |
| Trif | Glu | Glu | 53 | 24 | 5 | 0 |
| Trif | Gly+D | Gly+D | 96 | 98 | 9 | 959 |
| Trif | Gly+D+S | Gly+D | 97 | 99 | 8 | 947 |
| Trif | Gly+D | Glu | 73 | 68 | 9 | 0 |
| Trif | Gly+D | Glu+D | 94 | 94 | 3 | 728 |
| Trif | Gly+D | Glu+D+S | 96 | 97 | 3 | 1086 |
| Trif | S | Glu+D | 99 | 100 | 0 | 953 |
| Trif | Glu+D | Gly+D | 97 | 99 | 4 | 976 |
| Trif | Glu+D+S | Gly+D | 98 | 100 | 0 | 830 |
| Trif | Glu+D | Glu+D | 96 | 95 | 3 | 850 |
| Trif | S+Glu | Gly+Glu | 64 | 35 | 3 | 196 |
| Trif | S+Glu | Gly+D | 83 | 89 | 19 | 776 |
| None | None | None | 0 | 0 | 0 | 0 |
| LSD (0.05) | | | 7 | 10 | 7 | 325 |

[a]Herbicides abbreviations and doses: Trif, trifluralin at 1.12 kg ha$^{-1}$; Gly, glyphosate at 1.36 kg ae ha$^{-1}$; Glu, glufosinate at 0.59 kg ai ha$^{-1}$; Gly + D, glyphosate at 0.48 kg ha$^{-1}$ + 2,4-D choline at 0.45 kg ae ha$^{-1}$; S, *S*-metolachlor at 1.08 kg ai ha$^{-1}$; D, 2,4-D choline at 1.06 kg ae ha$^{-1}$.

**Table 12.** Palmer amaranth control with herbicide systems using glyphosate plus 2,4-D choline.

Cotton injury was greatest (19%) with trifluralin-applied PPI followed by *S*-metochlor plus glufosinate-applied EPOST followed by glyphosate plus 2,4-D choline applied MPOST (**Table 12**). Cotton lint yields were greatest with herbicide treatments which provided greater than 90% Palmer amaranth control with the exception of trifluralin applied PPI followed by glyphosate plus 2,4-D choline applied EPOST and MPOST.

In the glyphosate herbicide systems study, a late season rating suggests that the herbicide system which included glyphosate plus the pre-mix of rimsulfuron plus thifensulfuron-methyl applied preplant controlled Palmer amaranth less than 70%, while all other systems which included glyphosate plus either flumioxazin, fomesafen, or diruron applied preplant provided 89–99% control (**Table 13**). Diverse herbicide programs for controlling resistant Palmer amaranth and common waterhemp is an important herbicide-resistant management strategy [122]. Additionally, full labeled-use doses should always be used to achieve the greatest level

| Herbicide and application timing[a] | | | | | | |
|---|---|---|---|---|---|---|
| **PPI** | **PRE** | **EPOST** | **MPOST** | **LPOST** | **Control[b]** | **Yield** |
| | | | | | % | Kg ha$^{-1}$ |
| Gly+Flumi | Flume+Par | Gly + Ace | Gly+Ace | D+MSMA | 89 | 506 |
| Gly+Ace | Flume+Par | Gly + S | Gly+S | D+MSMA | 93 | 794 |
| Gly+Ace | Ace+Flume+Par | Glu + Pyr | Gly | D+MSMA | 99 | 801 |
| Gly+Rim+Thi | Pyr+Par | Gly + Dim | Gly+Dim | D+MSMA | 68 | 609 |
| Gly+D | Ace+Par | Gly + Pyr | Gly+S | D+MSMA | 91 | 770 |
| Gly+Flumi | Ace+Flume+Par | Gly + Ace | Gly | D+MSMA | 89 | 753 |
| None | None | None | None | None | 0 | 0 |
| LSD (0.05) | | | | | 16 | 274 |

[a]Herbicide abbreviations and doses: Ace, acetochlor at 1.27 kg ai ha$^{-1}$; Dim, dimethenamid-P at 0.63 kg ai ha$^{-1}$; D, direx at 1.12 kg ai ha$^{-1}$; Flumi, flumioxazin at 0.07 kg ai ha$^{-1}$; Flume, flumeturon at 1.12 kg ai ha$^{-1}$; Fome, fomesafen at 0.28 kg ai ha-1; Gly, glyphosate at 1.3 kg ae ha$^{-1}$; Gly + S, a premix of glyphosate at 0.95 kg ae ha$^{-1}$ + *S*-metochlor at 1.26 kg ha$^{-1}$; MSMA at 2.11 kg ai ha$^{-1}$; Par, paraquat at 0.56 kg ai ha$^{-1}$; Pyr, pyrithiobac at 0.06 kg ai ha$^{-1}$; Rim + Thi, a premix of rimsulfuron at 0.02 kg ha$^{-1}$ + thifensulfuron-methyl at 0.04 kg ha$^{-1}$; MSMA, MSMA at 2.11 kg ai ha$^{-1}$.

[b]Palmer amaranth control and cotton yield combined over years (2015, 2016) due to lack of year by treatment interaction.

**Table 13.** Palmer amaranth control and cotton response to herbicide systems.

of possible control and reduce the likelihood for the evolution of resistance. These results further displayed the high level of weed control this new technology is capable of providing. Also, emphasis should be placed on a zero-tolerance weed threshold [17], herbicides should also be applied at or less than the recommended weed height, and programs should not begin with an EPOST or MPOST application but rather start prior to planting with the application of residual herbicides.

While a few late emerging Palmer amaranth plants may be considered as being harmless, previous research has reported that late season Palmer amaranth seedlings are capable of seed production within 30 days after emergence [123]. Previous research also has shown that weeds left in the field at the time of harvest have the potential to enter harvesting machinery and be distributed across the field [124]. Thus, leaving weeds in the field prior to harvest can result in spreading viable weed seeds across the field. This practice will not only lead to increasing weed populations in that field but will also negatively impact sustainable weed management [125].

Also, a major challenge in managing weeds is minimizing the return of weed seed to the soil seed bank [126]. Menges [127] reported that maintaining fields weed free for 6 years reduced the soil seed bank of Palmer amaranth by 98%; however, 18 million seeds ha$^{-1}$ remained in the soil. Palmer amaranth seed viability decreased when buried below the depth of optimal germination for at least 36 months [128]. Given that Palmer amaranth has become the most challenging weed to manage in corn and cotton [129], understanding population dynamics of this weed may help lead to strategies that more effectively manage this weed.

## Author details

W. James Grichar[1]*, Joshua A. McGinty[1], Peter A. Dotray[2] and Travis W. Janak[3]

*Address all correspondence to: w-grichar@tamu.edu

1 Texas A&M AgriLife Research and Extension Center, Corpus Christi, Texas, USA

2 Texas A&M AgriLife Research and Extension Center, Lubbock, Texas, USA

3 B-H Genetics, Ganado, Texas, USA

## References

[1] Wiggins MS, McClure MA, Hayes R, Steckel LE. Integrating cover crops and POST herbicides for glyphosate-resistant Palmer amaranth (*Amaranthus palmeri*) control in corn. Weed Technology. 2015;**29**:412-418

[2] Anonymous, Monsanto Biotechnology Trait Acres: Fiscal Years 1996-2009 [Internet]. 2009. Available from: http://www.monsanto.com/investors/documents/2009/q4.biotech.acres

[3] Culpepper AS. Glyphosate-induced weed shifts. Weed Technology. 2006;**20**:277-281

[4] Owen MDK. Weed species shifts in glyphosate-resistant crops. Pest Management Science. 2008;**64**:377-387

[5] Culpepper AS, York AC. Weed management in glyphosate tolerant cotton. Journal of Cotton Science. 1998;**2**:174-185

[6] Johnson WG, Bradley PR, Hart SE, Buesinger ML, Massey RE. Efficacy and economics of weed management in glyphosate-resistant corn (*Zea mays*). Weed Technology. 2000;**14**:578-585

[7] Burke IC, Thomas WE, Allen JR, Collins J, Wilcut JW. A comparison of weed control in herbicide-resistant, herbicide-tolerant, and conventional corn. Weed Technology. 2008;**22**:571-579

[8] Bradley PR, Johnson WG, Hart SE, Buesinger ML, Massey RE. Economics of weed management in glufosinate-resistant corn (*Zea mays* L.). Weed Technology. 2000;**14**:495-501

[9] Ritter RL, Menbere H. Weed management systems utilizing glufosinate-resistant corn (*Zea mays*) and soybean (*Glycine max*). Weed Technology. 2001;**15**:89-94

[10] Heap I. The international survey of herbicide resistant weeds [Internet]. 2014. Available from: http://www.weedscience.com.pdf [Accessed: December 8, 2016]

[11] Morgan G, Baumann P, Chandler J. Competitive impact of Palmer amaranth (*Amaranthus palmeri*) on cotton (*Gossypium hirsutum*) development and yield. Weed Technology. 2001;**15**:408-412

[12] Smith D, Baker R, Steele D. Palmer amaranth (*Amaranthus palmeri*) impacts on yield, harvesting, and ginning in dryland cotton (*Gossypium hirsutum*). Weed Technology. 2000;**14**:122-126

[13] Stephenson DO, Bond IV JA, Landry RL, Edwards HM. Weed management in corn with postemergence applications of tembotrione or thiencarbazone: Tembotrione. Weed Technology. 2015;**29**:350-358

[14] Bollman JD, Boerboom CM, Becker RL, Fritz VA. Efficacy and tolerance to HPPD-inhibiting herbicides in sweet corn. Weed Technology. 2008;**20**:267-274

[15] Stephenson DO IV, Bond JA. Evaluation of thiencarbazone-methyl- and isoxaflutole-based herbicide programs in corn. Weed Technology. 2012;**26**:37-42

[16] Walsh MJ, Stratford K, Stone K, Powles SB. Synergistic effects of atrazine and mesotrione on susceptible and resistant wild radish (*Raphanus rephanistrum*) populations and the potential for overcoming resistance to triazine herbicides. Weed Technology. 2012;**26**:341-347

[17] Norsworthy JK, Griffith GM, Scott RC, Smith KL, Oliver LR. Confirmation and control of glyphosate-resistant Palmer amaranth (*Amaranthus palmeri*) in Arkansas. Weed Technology. 2008;**22**:108-113

[18] Hayes RM, Mueller TC, Willis JB, Montgomery RF. Glyphosate-resistant horseweed and factors influencing its control. Proceedings of the Southern Weed Science Society. 2002;**55**:119-120

[19] Yancy CH. Glyphosate Resistant Horseweed Causing Concern in North Carolina. Southeast Farm Press [Internet]. Available from: http://www.southeastfarmpress.com/glyphosate-resistant-horseweed-causing-concern-north-carolina [Accessed: January 11, 2017]

[20] Young B. Changes in herbicide use patterns and production practices resulting from glyphosate-resistant crops. Weed Technology. 2006;**20**:301-307

[21] Legleiter T, Bradley K. Glyphosate and multiple herbicide resistance in common waterhemp (*Amaranthus rudis*) populations from Missouri. Weed Science. 2008;**56**:582-587

[22] NRCS-National Resources Conservation Service. Plants Database: Amaranthus L. [Internet]. 2011. Available from: http://plants.usda.gov/java/profile?symbol=AMARA [Accessed: December 28, 2016]

[23] Van Wychen L. 2015 baseline survey of the most common and troublesome weeds in the United States and Canada. Weed Science Society of America National Weed Survey Dataset [Internet]. 2016. Available from: http://wssa.net/wp-content/Uploads/2015_Weed_Survey_Final.xlsx [Accessed: December 12, 2017]

[24] Elmore CD, editor. Weed Identification Guide. Champaign, IL: Southern Weed Science Society; 1985. p. 225

[25] Anonymous. Palmer amaranth–weed of TWO years-2014-2015 [Internet]. 2015. Available from: https://www.ag.ndsu.edu/weeds/weed-of-the-year-files/palmer-amaranth-2 [Accessed: December 3, 2016]

[26] Correll DS, Johnston MC. Manual of the Vascular Plants of Texas. 2nd ed. Richardson, TX: University Texas at Dallas; 1979. pp. 555-556

[27] Light GG, Mohammed MY, Dotray PA, Chandler JM, Wright RJ. Glyphosate-resistant common waterhemp (*Amatanthus rudis*) confirmed in Texas. Weed Technology. 2011;**25**: 480-485

[28] Franssen AS, Skinner DZ, Al-Khatib K, Horak MJ, Kulakow PA. Interspecific hybridization and gene flow of ALS-resistance in Amaranthus species. Weed Science. 2001;**49**: 598-606

[29] Sauer JD. Recent migration and evolution of the dioecious amaranths. Evolution. 1957;**11**: 11-31

[30] Sellers BA, Smeda RJ, Johnson WG, Kendig JA, Ellersieck MR. Comparative growth of six Amaranthus species in Missouri. Weed Science. 2003;**51**:329-333

[31] Bond JA, Oliver LR. Comparative growth of Palmer amaranth (*Amaranthus palmeri*) accessions. Weed Science. 2006;**54**:121-126

[32] Horak MJ, Loughin TM. Growth analysis of four Amaranthus species. Weed Science. 2000;**48**:347-355

[33] Keeley PE, Carter CH, Thullen RJ. Influence of planting date on growth of Palmer amaranth (*Amaranthus palmeri*). Weed Science. 1987;**35**:199-204

[34] Franssen AS, Skinner DZ, Al-Khatib K, Horak MJ. Pollen morphological differences in Amaranthus species and interspecific hybrids. Weed Science. 2001;**49**:732-737

[35] Guo P, Al-Khatib K. Temperature effects on germination and growth of redroot pigweed (*Amaranthus retroflexus*), Palmer amaranth (*A. palmeri*), and common waterhemp (*A. rudis*). Weed Science. 2003;**51**:329-333

[36] Vyn JD, Swanton CJ, Weaver SE, Sikkema PH. Control of *Amaranthus tuberculatus* var. *rudis* (common waterhemp) with pre- and postemergence herbicides in Zea mays L. Crop Protection. 2006;**35**:1051-1056

[37] Geier PW, Stahlman PW. Efficacy of isoxaflutole alone and in combination in corn. Proceedings of the North Central Weed Science Society. 1997;**52**:81

[38] Armel GR, Wilson HP, Richardson RJ, Hines TE. Mesotrione, acetochlor, and atrazine for weed management in corn (*Zea mays*). Weed Technology. 2003;**17**:284-290

[39] Armel GR, Wilson HP, Richardson RJ, Hines TE. Mesotrione combinations in no-till corn (*Zea mays*). Weed Technology. 2003;**17**:111-116

[40] Wilcut JW, Jordan DL, Vencill WK, Richburg III JS. Weed management in cotton (*Gossypium hirsutum*) with soil-applied and post-directed herbicides. Weed Technology. 1997;**11**:221-226

[41] Welch AK, Rahn PR, Voth RD, Mills JA, Shumway CR. Evaluation of preplant and preemergence herbicides in Roundup Ready® cotton. Proceedings of the Beltwide Cotton Conferences. 1997;**21**:784

[42] Gullickson G. 6 Points to Remember about Atrazine [Internet]. 2012. Available from: Agriculture.com [Accessed: January 10, 2017]

[43] Sutton P, Richards C, Buren L, Glasgow L. Activity of mesotrione on resistant weeds in maize. Pest Management Science. 2002;**58**:981-984

[44] Swanton CJ, Gulden RH, Chandler K. A rationale for atrazine stewardship in corn. Weed Science. 2007;**55**:75-81

[45] Tapia LS, Bauman TT, Harvey RG, Kells JJ, Kapusta G, Loux MM, et al. Postemergence herbicide application timing effects on annual grass control and corn (*Zea mays*) grain yield. Weed Science. 1997;**45**:138-143

[46] Baker JL, Johnson HP. The effects of tillage systems on pesticides in runoff from small watersheds. Transactions of the American Society of Agricultural Engineers. 1979;**22**:554-559

[47] Pantone DJ, Young RA, Buhler DD, Eberlein CV, Koskinen WC, Forcella F. Water quality impacts associated with pre- and postemergence applications of atrazine in maize. Journal of Environmental Quality. 1992;**21**:567-573

[48] Shaner D, editor. Herbicide Handbook. 10th ed. Lawrence, KS: Weed Science Society of America; 2014. p. 513

[49] Behrens M, Mutlu N, Chakraborty S, Dumitru R, Jiang W, LaVallee B, et al. Dicamba resistance: Enlarging and preserving biotechnology-based weed management strategies. Science. 2007;**316**:1185-1188

[50] USDA-APHIS. Determination of Nonregulated Status for Monsanto Company MON 88701 cotton. United States Department of Agriculture. Animal and Plant Health Inspection Service. APHIS-2013-0043-4809 [Internet]. Available from: https://www.aphis.usda.gov/brs/aphisdocs/12_18501p_det.pdf [Accessed: January 15, 2017].

[51] USDA-AMS. Cotton Varieties Planted 2016 Crop. Memphis, TN: United States Department of Agriculture. Agricultural Marketing Service – Cotton and Tobacco Program. 2016. p. 10

[52] Richburg JS, Wright JR, Braxton LB, Robinson AE. Inventors; Dow AgroSciences, Assignee. Increased Tolerance of DHT-Enabled Plants to Auxinic Herbicides Resulting from Moiety Differences in Auxinic Molecule Structures. US patent 13,345,236

[53] Peterson GE. The discovery and development of 2,4-D. Agricultural History. 1967;**41**:243-254

[54] Kruger GR, Davis VM, Weller SC, Johnson WG. Response and survival of rosette-stage horseweed (*Conyza canadensis*) after exposure to 2,4-D. Weed Science. 2008;**56**:748-752

[55] Robinson AP, Simpson DM, Johnson WG. Summer annual weed control with 2,4-D and glyphosate. Weed Technology. 2012;**26**:657-660

[56] Siebert JD, Griffin JL, Jones CA. Red morningglory (*Ipomoea coccinea*) control with 2,4-D and alternative herbicides. Weed Technology. 2004;**18**:38-44

[57] White RH, Worsham AD. Control of legume cover crops in no-till corn (*Zea mays*) and cotton (*Gossypium hirsutum*). Weed Technology. 1990;**4**:57-62

[58] Egan JF, Barlow KM, Mortensen DA. A meta-analysis on the effects of 2,4-D and dicamba drift on soybean and cotton. Weed Science. 2014;**62**:193-206

[59] Mortensen DA, Egan JF, Maxwell BD, Ryan MR, Smith RG. Navigating a critical juncture for sustainable weed management. Bioscience. 2012;**62**:75-85

[60] Riar DS, Norsworthy JK, Steckel LE, Stepehenson IV DO, Bond JA. Consultant perspectives on weed management needs in Midsouthern United States cotton: A follow-up survey. Weed Technology. 2013;**27**:778-787

[61] Abel S. Dow AgroSciences Company Petition for Determination of Nonregulated Status of 2,4-D and Glufosinate-Resistant DAS-81910-7 Cotton. USDA Environmental Assessment 2015 [Internet]. Available from: https://www.aphis.usda.gov/brs/aphisdocs/13_26201p_fea.pdf [Accessed: January 15, 2017]

[62] Staten G. Contamination of cotton fields by 2,4-D or hormone-type weed sprays. Agronomy Journal. 1946;**38**:536-544

[63] Everitt JD, Keeling JW. Cotton growth and yield response to simulated 2,4-D and dicamba drift. Weed Technology. 2009;**23**:503-506

[64] Johnson VA, Fisher LR, Jordan DL, Edminsten KE, Stewart AM, York AC. Cotton, peanut, and soybean response to sublethal rates of dicamba, glufosinate, and 2,4-D. Weed Technology. 2012;**26**:195-206

[65] Sciumbato AS, Chandler JM, Senseman SA, Bovey RW, Smith KL. Determining exposure to auxin-like herbicides. I. Quantifying injury to cotton and soybean. Weed Technology. 2004;**18**:1125-1134

[66] Beckie HJ. Herbicide-resistant weeds: Management tactics and practices. Weed Technology. 2006;**20**:793-814

[67] Putwain PD. Herbicide resistance in weeds-an inevitable consequence of herbicide use? Proceedings Brighton Crop Protection Conference—Weeds. Farnham, UK: British Crop Protection Council. 1982. pp. 719-728

[68] Wrubel RP, Gressel J. Are herbicide mixtures useful for delaying the rapid evolution of resistance? A case study. Weed Technology. 1994;**8**:635-648

[69] Beckie HJ, Hall LM, Tardif FJ. Impact and management of herbicide-resistant weeds in Canada. Proceedings Brighton Crop Protection Conference—Weeds. Farnham, UK: British Crop Protection Council; 2001. pp. 747-754

[70] Diggle AJ, Neve PB, Smith FP. Herbicides used in combination can reduce the probability of herbicide resistance in finite weed populations. Weed Research. 2003;**43**:371-382

[71] SAS Institute Incorporated. SAS/STAT User's Guide: Statistics, Version 9.1. Cary, NC, USA: SAS Institute; 2007. p. 204

[72] Anonymous. Crop Protection Reference. 29th ed. Chemical & Pharmaceutical Press, Inc. New York [Internet]. 2014. Available from: http://www.greenbook.net [Accessed: December 29, 2016]

[73] Buchanan GA, Murray DS, Hauser EW. Weeds and their control in peanuts. In: Pattee HE, Young CT, editors. Peanut Science and Technology. Yoakum, TX: American Peanut Research Education Society; 1982. pp. 209-249

[74] Chamblee RW, Thompson L Jr, Bunn TM. Management of broadleaf signalgrass (*Brachiaria platyphylla*) in peanuts (*Arachis hypogaea*). Weed Science. 1982;**30**:40-44

[75] Wilcut JW, Wehtje GR, Patterson MG. Economic assessment of weed control systems for peanuts (*Arachis hypogaea*). Weed Science. 1987;**35**:433-437

[76] Wilcut JW, York AC, Wehtje GR. The control and interaction of weeds in peanut (*Arachis hypogaea*). Reviews of Weed Science. 1994;**6**:177-205

[77] Wilcut JW, York AC, Grichar WJ, Wehtje GR. The biology and management of weeds in peanut (*Arachis hypogaea*). In: Pattee HE, Stalker HT, editors. Advances in Peanut Science. Stillwater, OK: American Peanut Research Education Society; 1995. pp. 207-244

[78] Ross MA, Childs DJ. Herbicide mode of action summary. WS-23-W [Internet]. 1996. Available from: http://www.extension.purdue.edu/extmedia/ws/ws-23-w.html [Accessed: January 14, 2017]

[79] Grichar WJ, Colburn AE, Keraney NS. Herbicides for reduced tillage production in peanut (*Arachis hypogaea*) in the Southwest. Weed Technology. 1994;**8**:212-216

[80] Steele GL, Porpiglia PJ, Chandler JM. Efficacy of KIH-485 on Texas panicum (*Panicum texanum*) and selected broadleaf weeds in corn. Weed Technology. 2005;**19**:866-869

[81] Anonymous. KIH-485 Herbicide. Technical Information. White Plains, NY: K-I Chemical USA, Inc., 2003. p. 6

[82] Ohmes GA, Kendig JA, Barham RL, Ezell PM. Efficacy of ZA1296 in corn. Proceedings of the Southern Weed Science Society. 2000;**53**:225

[83] Allen JR, Johnson WG, Smeda RJ, Kremer RJ. ALS-resistant Helianthus annuus interference in *Glycine max*. Weed Science. 2000;**48**:461-466

[84] Al-Khatib K, Peterson DE, Regehr DL. Control of imazethapyr-resistant common sunflower (*Helianthus annuus*) in soybean (*Glycine max*) and corn (*Zea mays*). Weed Technology. 2000;**14**:133-139

[85] Johnson WG, Chahal GS, Regehr DL. Efficacy of various corn herbicides applied preplant incorporated and preemergence. Weed Technology. 2012;**26**:220-229

[86] Taylor-Lovell S, Wax LM. Weed control in field corn (*Zea mays*) with RPA 201772 combinations with atrazine and S-metolachlor. Weed Technology. 2011;**15**:249-256

[87] Sprague CL, Kells JJ, Penner D. Effect of application timing on corn tolerance and weed control with isoxaflutole. Weed Science Society of America Abstracts. 1997;**37**:13

[88] Prostko EP, Grey TL, Davis JW. Texas panicum (*Panicum texanum*) control in irrigated field corn (*Zea mays*) with foramsulfuron, glyphosate, nicosulfuron, and pendimethalin. Weed Technology. 2006;**20**:961-964

[89] Lamore D, Simkins G, Watteyne K, Allen J. Weed control programs with tembotrione in corn. Proceedings of the North Central Weed Science Society. 2006;**61**:119

[90] Dalley CD, Bernards ML, Kells JJ. Effect of weed removal and row spacing on soil moisture in corn (*Zea mays*). Weed Technology. 2006;**20**:399-409

[91] Gower SA, Loux MM, Cardina J, Harrison SK, Sprankle PL, Probst NJ, et al. Effect of postemergence glyphosate application timing on weed control and grain yield in glyphosate-resistant corn: Results of a 2-yr multistate study. Weed Technology. 2003;**17**:821-828

[92] Norsworthy JK, Burgos NR, Oliver LR. Differences in weed tolerance to glyphosate involve different mechanisms. Weed Technology. 2001;**15**:725-731

[93] Shaw DR, Arnold JC. Weed control from herbicide combinations with glyphosate. Weed Technology. 2002;**16**:1-6

[94] Starke RJ, Oliver LR. Interaction of glyphosate with chlorimuron, fomesafen, imazethapyr, and sulfentrazone. Weed Science. 1998;**46**:652-660

[95] Culpepper AS, Gimenez AE, York AC, Batts RB, Wilcut JW. Morningglory (*Ipomoea* spp.) and large crabgrass (*Digitaria sanguinalis*) control with glyphosate and 2,4-DB mixtures in glyphosate resistant soybean (*Glycine max*). Weed Technology. 2001;**15**:56-61

[96] Norsworthy JK, Oliver LR. Effect of irrigation, soybean (*Glycine max*) density, and glyphosate on hemp sesbania (*Sesbania exaltata*) and pitted morningglory (*Ipomoea lacunosa*) interference with soybean. Weed Technology. 2002;**16**:7-17

[97] Reedy KN, Whiting K. Weed control and economic comparisons of glyphosate-resistant, sulfonylurea-tolerant, and conventional soybean (*Glycine max*) systems. Weed Technology. 2000;**14**:204-211

[98] Bararpour MT, Oliver LR, Bell CG. Comparison of HPPD inhibitors for weed control programs in corn. Proceedings of the Southern Weed Science Society. 2011;**64**:16

[99] Grier PW, Stahlman PW. EXP 31130A efficacy and corn (*Zea mays*) response in western Kansas. Weed Technology. 1999;**13**:404-410

[100] Knezevic SZ, Sikkema PH, Tardif F, Hamill AS, Chandler K, Swanton CJ. Biologically effective dose and selectivity of isoxaflutole for preemergence weed control in corn (*Zea mays* L.). Weed Technology. 1998;**12**:670-676

[101] O'Sullivan J, Thomas RJ, Sikkema PH. Sweet corn (*Zea mays*) cultivar sensitivity to RPA 201772. Weed Technology. 2001;**15**:332-336

[102] Sikkema PH, Knezevic SZ, Hamil AS, Tardif F, Swanton CJ. Biologically effective dose and selectivity for SAN 1269H for weed control in corn (*Zea mays* L.). Weed Technology. 1999;**13**:283-289

[103] Sprague CL, Kells JJ, Penner D. Weed control and corn (*Zea mays*) tolerance from soil-applied RPA 201772. Weed Technology. 1999;**13**:713-725

[104] Sprague CL, Stoller EW, Hart SE. Preemergence broadleaf weed control and crop tolerance to imidazolinone-resistant and -susceptible corn (*Zea mays*). Weed Technology. 1997;**11**:118-122

[105] Sprague CL, Penner D. Basis for different corn tolerance of four corn hybrids to isoxaflutole. Proceedings of the North Central Weed Science Society. 1998;**53**:94

[106] Gower SA, Loux MM, Cardina J, Harrison SK. Effect of planting date, residual herbicide, and postemergence application timing on weed corn and grain yield in glyphosate-tolerant corn (*Zea mays*). Weed Technology. 2002;**16**:488-494

[107] Halford C, Hamill AS, Zhang J, Doucet C. Critical period of weed control in no-till soybean (*Glycine max*) and corn (*Zea mays*). Weed Technology. 2001;**15**:737-744

[108] Myers MW, Curran WS, Vangessel MJ, Majek BA, Scott BA, Mortensen DA, et al. The effect of weed density and application timing on weed control and corn grain yield. Weed Technology. 2005;**19**:102-107

[109] Dotray P, Keeling W, Henniger G, Abernathy J. Palmer amaranth (*Amaranthus palmeri*) and devil's claw (*Proboscidea louisianica*) control in cotton with pyrithiobac. Weed Technology. 1996;**10**:7-12

[110] Branson J, Smith K, Barrentine J. Comparison of trifloxysulfuron and pyrithiobac in glyphosate-resistant and bromoxynil-resistant cotton. Weed Technology. 2005;**19**:404-410

[111] Anonymous. Staple LX Herbicide Product Label [Internet]. 2011. Available from: http://www.cdms.net/LDat/Id7DD009.pdf [Accessed: January 12, 2017]

[112] Anonymous. Prowl $H_2O$ Herbicide Product Label [Internet]. 2016. Available from: http://www.cdms.net/Idat/Id6CT004.pdf [Accessed: January 12, 2017]

[113] Robison L, Fenster C. Residual effects of EPTC and trifluralin incorporated with different implements. Weed Science. 1968;**16**:415-417

[114] Wiese A, Chenault W, Hudspeth E. Incorporation of preplant herbicides for cotton. Weed Science. 1969;**17**:481-483

[115] Anonymous. Liberty 280 SL Herbicide Product Label [Internet]. Available from: https://www.cropscience.bayer.us/Products/herbicides/liberty/label-msds [Accessed: January 12, 2017]

[116] Craigmyle B, Ellis J, Bradley K. Influence of weed height and glufosinate plus 2,4-D combinations on weed control in soybean with resistance to 2,4-D. Weed Technology. 2013;**27**:271-280

[117] Powles S. Evolved glyphosate-resistant weeds around the world: Lessons to be learnt. Pest Management Science. 2008;**64**:360-365

[118] Evans J, Tranel P, Hager A, Schutte B, Wu C, Chatham L, Davis A. Managing the evolution of herbicide resistance. Pest Management Science. 2015;**72**:74-80

[119] Chahal GS, Johnson WG. Influence of glyphosate or glufosinate combinations with growth regulator herbicides and other agrochemicals in controlling glyphosate-resistant weeds. Weed Technology. 2012;**26**:638-643

[120] Miller MR, Norsworthy JK. Evaluation of herbicide programs for use in a 2,4-D- resistant soybean technology for control of glyphosate-resistant Palmer amaranth (*Amaranthus palmeri*). Weed Technology. 2016;**30**:366-376

[121] Riar DS, Norsworthy JK, Steckel LE, Stephenson DO, Eubank TW, Scott RC. Assessment of weed management practices and problematic weeds in the Midsouth United States-soybean: A consultant's perspective. Weed Technology. 2013;**27**:612-622

[122] Norsworthy JK, Ward SM, Shaw DR, Llewellyn RS, Nichols RL, Webster TM, et al. Reducing the risks of herbicide resistance: Best management practices and recommendations. Weed Science (Special Issue). 2012;**60**:31-62

[123] Jha P, Norsworthy JK. Soybean canopy and tillage effects on emergence of Palmer amaranth (*Amaranthus palmeri*) from a natural seed bank. Weed Science. 2009;**57**:644-651

[124] Walsh MJ, Powles SB. High seed retention at maturity of annual weeds infesting field crops highlights the potential for harvest weed seed control. Weed Technology. 2014;**28**:486-493

[125] Norsworthy JK, Walsh MJ, Bagavathiannan MV, Bradley KW, Steckel L, Kruger G, et al. Harvest weed seed control: Testing Australian seedbank management tactics in USA. Proceedings of the Weed Science Society of America. 2014;**250**

[126] Swanton CJ, Booth BD. Management of weed seedbanks in the context of populations and communities. Weed Technology. 2004;**18**:1496-1502

[127] Menges RM. Weed seed population dynamics during six years of weed management systems in crop rotations on irrigated soil. Weed Science. 1987;**35**:328-332

[128] Sosnoskie LM, Webster TM, Culpepper AS. Glyphosate resistance does not affect Palmer amaranth (*Amaranthus palmeri*) seedbank longevity. Weed Science. 2013;**61**:283-288

[129] Webster TM. Weed survey-southern states: Broadleaf crop subsection. Houston: Proceedings of the Southern Weed Science Society; 2013. pp. 275-287

Chapter 5

# Biochemical and Molecular Knowledge about Developing Herbicide-Resistant Weeds

Mohammad Taghi Alebrahim,
Rouzbeh Zangoueinejad and Te Ming Tseng

Additional information is available at the end of the chapter

http://dx.doi.org/10.5772/intechopen.69211

**Abstract**

Herbicide resistance is the genetic capacity of a weed population to survive an herbicide treatment that, under normal use conditions, would effectively control the resistant weed population. Weeds have been evolving in conventional crop cultivars worldwide from selection pressure placed on them from repeated use of herbicides. In this chapter, we intend to explain the biochemical and molecular basis of herbicide resistance in weeds. On the other hand, herbicide resistance can be a useful tool so that weed scientists can use as important approach to control and manage weeds. There are several strategies for the production of HR crops by genetic engineering and the methods used in this process will be discussed in this chapter.

**Keywords:** herbicide resistance, biochemical mechanisms, molecular basis

## 1. Introduction

Humans have travelled a long way reaching the agriculutre that is there today. In the initial days, weed control has been a major concern in crop production and different approaches have been tested to manage weeds. Some approaches have been retired after several years and others still being adopted. Herbicide application is one of the approaches that still remain durable and efficient. Similarly, for every Human-made strategy, herbicide application has both positive and negative effects. Herbicides have increased agricultural productivity effectively, but on the other hand, has caused a serious problem by promoting the evolution of herbicide-resistant weeds. Successive applications of some herbicides of the same group or some herbicides with the same mode of action in a field will contribute resistance to herbicides in one or several weed species. In spite of some concerns about weed resistance to herbicides, only a logical approach integrates all common strategies to inhabit herbicide resistance in weeds because the Human population is ever increasing.

Although development of resistance in weeds is an undesirable phenomenon, herbicide tolerance in crops is favorable. If the principle crop is not always tolerant to the herbicide, the herbicide will either decrease the productiveness of the primary crop or kill it. If the herbicide is not strong enough, it could allow the proliferation of weeds within the crop field thus affecting the productiveness of the primary crop. It is therefore desirable to produce crops that are tolerant to herbicides. The important objectives of this chapter are to clearly explain the important biochemical and molecular reasons of herbicide resistance in weeds, and and at the same time investigate the methods for the production of HR crops.

## 2. Definitions of tolerance and resistance

Generally, herbicide has very beneficial effects on agricultural production worldwide [1]. Herbicides are often the most reliable and economical option available to control weeds [2, 3]. The availability of herbicide has allowed that researchers modify plant height and transform plants for increased performance [4]. Efficiency and cost-effectiveness of herbicides has led to positive impact on the agricultural production systems in the developed countries [5]. Herbicide tolerance and herbicide resistance are two very important concepts that should be carefully considered. Standard definitions of the herbicide "tolerance" and "resistance" based on the crop and weed biology were established by the Weed Science Society of America (WSSA) in 1998. According to the definitions of the WSSA, tolerance is the inborn capacity of plant groups to survive and recreate after herbicide treatment. This infers there was no election or genetic manipulation to make the plant tolerant; it is naturally tolerant. Resistance is "the acquired capacity of a plant to survive and propagate after introduction to a dosage of herbicide typically deadly to the wild sort. Resistance may be innately happening or initiated by such strategies as genetic engineering or election of variations created by tissue culture or mutagenesis [6].

## 3. Herbicide resistance mechanisms

Fundamentally, two types of mechanisms are involved in resistance. Target-site resistance (TSR) is caused by changes in the tridimensional structure of the herbicide target protein that decreases herbicide binding, or by increased activity of the target protein. TSR is conferred by gene mutations in target enzymes such as 5-enolpyruvylshikimate-3-phosphate synthase, which is reported in many resistant weed species [7–9]. Non-target-site resistance (NTSR) is endowed by any mechanism not belonging to TSR, e.g., reduction in herbicide uptake or translocation in the plant, or enhanced herbicide detoxification [8, 10]. Mutations endowing herbicide resistance can be classified into two types. The first type is structural changes in a

DNA sequence encoding a protein, i.e., structural mutations. Structural mutations endowing herbicide resistance are expected to cause a structural modification in the tridimensional structure of a protein that will lead to a decrease in the efficacy of an herbicide. For example, mutations conferring an amino acid substitution at the herbicide-binding site of a target protein can decrease the affinity of the herbicide for the target protein (TSR). Alternatively, mutations at the active site of a metabolic enzyme or a transporter protein can improve the activity of these proteins in herbicide degradation or compartmentation away from its site of action, respectively (NTSR). In the case of structural changes in DNA sequence, seeking the cause for resistance means identifying and being able to detect the relevant structural mutations in the DNA of resistant plants. The second type of mutations associated with herbicide resistance results in a difference in the expression of one or several genes in resistant plants compared to sensitive plants, i.e., regulatory mutations [11, 12]. These mutations are changes in a DNA sequence that can cause an increase in the expression of the herbicide target protein that compensates for the herbicide inhibitory action (TSR), or a variation in the expression of herbicide-metabolizing enzyme(s) or of transporter proteins that will lead to an increase in herbicide degradation or compartmentation away from its site of action, respectively (NTSR) [13]. Non-target-site resistance compared with target-site resistance is less investigated especially in broadleaf weed species. Non-target-site resistance may cause weeds evolve unforeseeable resistance to diverse herbicides of different modes of action [14].

### 3.1. Target-site resistance

#### *3.1.1. Resistance to protoporphyrinogen IX oxidase-inhibiting herbicides*

Protoporphyrinogen oxidase (Protox), the target site of the diphenylether herbicides, catalyzes the conversion of protoporphyrinogen to protoporphyrin IX in tetrapyrrole biosynthesis. Several herbicides including the diphenylethers and oxidiazoles inhibit PPO. Inhibition of Protox leads to the production of large quantities of free protoporphyrin IX in the cytoplasm, which causes photodynamic damage in the presence of light and oxygen [15]. Results of investigations with a resistant *Amaranthus tuberculatus* biotype have showed an unprecedented and unanticipated mutation in which resistance is endowed by an amino acid deletion. Presumably, chloroplastic and mitochondrial protoporphyrinogen oxidase encodes by the PPX2L gene in resistant *Amaranthus tuberculatus*, there is the lack of a 3-bp codon, bringing an elimination of glycine at position 210 [16]. It is the just reportage card of codon/amino acid omission presenting resistance to herbicide. The Gly-210 elimination in the protoporphyrinogen oxidase gene confers extremely rate resistance to protoporphyrinogen oxidase herbicides by minimal impact on the natural inclination of protoporphyrinogen oxidase for its substrate protogen; however, the omission causes 10-fold lower protoporphyrinogen oxidase activity toward the ferocious sort [17]. However, resistance to Protox inhibitors has been selected for cell cultures [18] and has been generated in transgenic plants expressing heterologous Protox genes [15]. For example, a Protox Val389 to meet substitution endowed resistance in a selected *Chlamydomonas reinhardtii* line [19]. It has been proposed that the introduced resistant forms of Protox would need to replace rather than simply supplying the endogenous plant enzyme in order to avoid production of the toxic oxygen species following herbicide treatment [15, 20].

An obvious question is whether Gly-210 substitution, rather than deletion, would endow resistance. Modeling demonstrated that substitutions at Gly-210 provide either little or no resistance, or greatly decrease PPO functionality [56]. The necessity for contemporary absence of three nucleotides in the encoding succession of the focus gene, in addition to the duplex focusing of the gene result should chloroplasts and mitochondria, ought to restrict the development about this omission resistance mechanism, however it has been demonstrated in a further four resistant *A. tuberculatus* societies [21].

#### *3.1.2. Resistance to tubulin assembly inhibiting herbicides*

Both target-site resistance and non-target-site resistance to tubulin herbicides exist [22]. Target-site-based resistance to dinitroaniline herbicides has evolved in several species, such as *Setaria viridis* and *Eleusine indica*. Dinitroaniline and other tubulin-inhibiting herbicides have been used for several decades, and evolved resistance has been reported in some weed species (only 12 weed species) [23]. The mode of action of this group of herbicides is to bind to plant tubulin dimers and disrupting microtubule growth [24, 25]. In fact, these herbicides inhibit cell division by binding to the tubulin monomers, preventing their polymerization and spindle fiber formation [24]. Microtubules are polymers of $\alpha$- and $\beta$-tubulin dimers and are involved in many essential cellular processes, including mitosis, cytokinesis, and vesicular transport [23]. Several possible resistance mechanisms have been proposed, including microtubule hyperstabilization and posttranslational modification [26, 27], but decisive document for these is still wanting. However, witness is beginning to stack up for target-site mutations. Analysis of resistant *Eleusine indica* biotypes has shown that a Thr239 to Ile mutation in a a-tubulin gene endows a high level of resistance, provided a Met268 to Thr mutation confers a lower or intermediate level of resistance [28, 29]. The Thr239 to Ile mutation conferred resistance to oryzalin, pendimethalin, and amiprophos-methyl, but not to pronamide in transgenic tobacco [30]. Similarly, Lys350 to Glu or Met mutations in a b-tubulin gene conferred resistance to colchicine (which also inhibits cell division) in *Chlamydomonas reinhardtii* [31]. However, the latter mutations have not been reported in any higher plant resistant to dinitroaniline herbicides [15].

#### *3.1.3. Resistance to 5-enolpyruvylshikimate-3-phosphate synthase inhibitors*

Glyphosate, a widely used nonselective herbicide, inhibits 5-enolpyruvylshikimate-3-phosphate synthase (EPSPS), a key enzyme in the biosynthesis of the aromatic amino acids phenylalanine and tyrosine [15]. Glyphosate inhibits potently the chloroplast enzyme 5-enolpyruvylshikimate-3-phosphate synthase (EPSPS), which catalyzes the reaction of shikimate-3-phosphate (S3P) and phosphoenolpyruvate (PEP) to form 5-enolpyruvylshikimate-3-phosphate (EPSP). Glyphosate blockage of EPSPS activity interrupts the shikimate pathway and prevents aromatic amino acid product, ultimately causing plant death [32].

A major factor accelerating the evolution of glyphosate-resistant weeds has been the advent of transgenic glyphosate-resistant crops, such as soybean, maize, cotton, and canola. In these crops, glyphosate has replaced almost all other herbicides or other means of achieving weed control. From an evolutionary viewpoint, this singular reliance on glyphosate is an intense selection for any glyphosate-resistance genes [33, 9]. A serine substitution at Pro-106 (Pro-106-Ser) in an extremely conserved district of the EPSPS gene creates target-site glyphosate resistance, which was first observed in an *Eleusine indica* biotype [34, 35]. After that, as a first report, in glyphosate-resistant *E. indica* and *Lolium* populations have been first detected threonine and alanine substitutions at Pro-106 [36, 37]. Pro-106 (Pro-106-Ser) amino acid substitutions have been recognized in *E. indica* and also *Lolium* populations around the globe. These Pro-106 substitutions confer only a modest degree of glyphosate resistance [38].

Some researchers suggest that the resistant population of *Lolium rigidum* presents three different mechanisms of resistance to glyphosate, namely reduced absorption, reduced mobility in the plants, and a mutation in the gene coding for the enzyme targeted by glyphosate [39]. The crystal structure of *Escherichia coli* EPSPS and molecular modeling displays that glyphosate barricades EPSPS by engrossing the PEP binding site [40, 41]. Based on results of a decisive study on *E. coli* EPSPS Pro-106 substitutions and the crystal structure of EPSPS-S3P-glyphosate, it was found that a little restricted of the glyphosate/PEP binding site hole is created by Pro-106 substitutions, which maintains EPSPS functionality but confers glyphosate resistance [40]. In comparison, high-level glyphosate resistance is conferred by substitutions at Gly-101 or Thr-102, which decreases the content of the glyphosate/PEP binding and also this significantly decreases dependence for PEP [42]. Therefore, the protecting of EPSPS functionality may be very scarce in mutations, empowering both glyphosate and PEP binding [43].

It was shown that more than 40-fold EPSPS overexpress as a result of up to 100-fold EPSPS gene amplification. There are some evidences that proofed this fact in highly glyphosate-resistant *Amaranthus palmeri* biotypes several years ago. This inheritable EPSPS gene amplification can affect the expression level and glyphosate resistance segregating in F2 generation plants [44]. Nowadays, some laboratory attempts is doing to protect this kind of field-evolved resistance by selection of glyphosate-resistant cell lines from several plant species with EPSPS gene amplification. For example, it was shown a three-fold increase in basal EPSPS mRNA and enzyme activity in glyphosate-resistant *L. rigidum*, and a supplementary higher EPSPS expression in some glyphosate-resistant *Conyza* biotypes. However, in these given species, glyphosate translocation decrease considers as the most important resistance mechanism [45, 46]. Breeders showed although EPSPS relative copy number in nuclear genome can positively influence EPSPS mRNA level, EPSPS protein amount and activity female parents have major role than male parents in transformation of resistance inheritance [47].

Multiple herbicide resistance evolves various heterologous resistance mechanisms enciphered by particular resistance genes that coexist at the individual and/or population level, conferring resistance to several herbicides with different modes of action. Given its significance in modern agriculture, the most serious multiple herbicide resistance scenarios are those involving glyphosate [48].

#### *3.1.4. Resistance to ACCase-inhibiting herbicides*

The aryloxyphenoxypropionate (AOPP) and cyclohexanedione (CHD) herbicides inhibit acetyl-CoA carboxylase (ACCase) [49, 50]. Two types of ACCase have been identified: the heteromeric prokaryotic ACCase is composed of multiple subunits, whereas the homomeric eukaryotic ACCase is a large multidomain protein. Thus, most dicot species tolerate ACCase-inhibiting herbicides well, but most grass species are susceptible, meaning that ACCase herbicides control only grass weed species [51]. Multiple forms of eukaryotic ACCase are present in some grasses, which differ in herbicide sensitivity [52]. This is the primary basis for selectivity of these herbicides between grasses and dicots. Some grass species, including some cereal crops, are tolerant of these herbicides based on their ability to metabolize the herbicides to inactive compounds [53]. In addition, some grasses are tolerant due to an insensitive form of ACCase [54, 55].

Hedgehog dogtail (*Cynosurus echinatus*) is an annual grass, native to Europe, additionally broadly conveyed in North and South America, South Africa, and Australia. Two hedgehog dogtail biotypes, one diclofop-methyl (DM) safe and one DM vulnerable, were examined in detail for exploratory measurements reaction resistance components. The digestion system of 14CDM, D-corrosive, and D-conjugate metabolites were recognized by thin-layer chromatography. The acetyl-CoA carboxylase *in vitro* tests demonstrated that the objective site was exceptionally touchy to aryloxyphenoxy propanoate, cyclohexanedione, and phenylpyrazoline herbicides in the *Cenchrus echinatus* susceptible biotype, provided the resistant biotype was coldhearted to the already specified herbicides. DNA sequencing concentrates affirmed that *Cenchrus echinatus* cross-imperviousness to acetyl-CoA carboxylase inhibitors has been presented by particular acetyl-CoA carboxylase two-fold point transformations Ile-2041-Asn and Cys-2088-Arg [48].

The results of enzyme inhibition studies suggest several distinct altered forms of ACCase associated with different levels of resistance to various ACCase inhibitors [56]. Mostly, resistance to aryloxyphenoxy propanoate and cyclohexanedione herbicides is owing to a mutation in the objective enzyme, making it lesser susceptible to blockage by these herbicides. The results of enzyme deterrence investigates propose some different modified figures of acetyl CoA carboxylase correlated with various measures of resistance to distinct acetyl CoA carboxylase inhibitors. In contrast, a second biotype was very resistant to sethoxydim (R/S I50 ratio of 420), but had only a low level of resistance to other AOPP and CHD herbicides [57]. A similar pattern was observed in a *Setaria faberi* biotype from Iowa and in a sethoxydim-resistant corn line selected in tissue culture [58, 57]. A third pattern of resistance, conferring high-level resistance to fluazifop and lower levels of resistance to other AOPP and CHDs, has been found in a biotype of *Eleusine indica* from Malaysia [59], a *Lolium rigidum* biotype from Australia [60]. In a fourth category, some biotypes are resistant to AOPP herbicides but not to CHDs. These include *L. rigidum* biotype a VLR69 from Australia [61], a *Lolium multiyorum* biotype from Oregon, the USA [62], and *Avena fatua* biotype UM33 from Manitoba, Canada [63]. Similar groupings of resistant biotypes have been proposed in Ref. [64] according to entire plant cross-resistance templates to aryloxyphenoxy propanoate and cyclohexanedione herbicides in *Avena fatua* biotypes from Canada. Because of the two different acetyl CoA carboxylase genes in the weed grasses, this plant family encodes both cytosolic and plastidic figures of the acetyl CoA carboxylase. The target form of acetyl CoA carboxylase for aryloxyphenoxy propanoate and cyclohexanedione herbicides is the plastidic form, and in fact, this form of acetyl CoA carboxylase is modified in resistant weed biotypes [65]. Generally, the different patterns of resistance may be endowed by separate mutations in the gene for plastidic ACCase. Some reports indicate that at least one of the mutations is located in the carboxyltransferase region, toward the C terminal end of plastidic ACCase [66, 67]. Further molecular analysis is required to confirm the identity of this and other mutations responsible for resistance to these herbicides [15].

### *3.1.5. Resistance to AHAS (ALS)-inhibiting herbicides*

Acetolactate synthase (ALS) is the first enzyme in the biosynthetic for the branched-chain amino acids, such as valine, leucine, and isoleucine. A large number of herbicides, for example, sulfonylurea (SU), imidazolinone (IMI), triazolopyrimidine, pyrimidinyl-thiobenzoates, and sulfonyl-aminocarbonyl-triazolinone effect on acetohydroxyacid synthase (AHAS) catalyzes the formation of both aceto-hydroxybutyrate and acetolactate [68]. The vast AHAS-inhibiting herbicide resistance literature has been thoroughly reviewed [69, 70], so here, we are focusing on last expansions. It was rapidly established that AHAS herbicide-resistant plants could have a mutant, resistant AHAS enzyme [71, 58], and reports of resistant AHAS in many weeds followed. At Pro-197, 11 amino acid substitutions can endow AHAS herbicide resistance [51]. Although faster herbicide detoxification is a mechanism in some biotypes, in most cases, resistance to ALS and AHAS herbicides is endowed by target-site mutations [11, 72]. Target-site-based ALS resistance is due to point mutations that occur within discrete conserved domains of the ALS gene [11]. Most resistance mutations occur at the Pro-197 position, including one based on a double mutation [32]. Pro-197 mutations confer a high level of resistance to sulfonylurea herbicides, but low or no cross-resistance to imidazolinone herbicides. The Trp591 to Leu mutation confers high levels of resistance to all ALS inhibitors, whereas the Ser670 to Asp and Ala122 to Thr mutations confer a high level of resistance to imidazolinones but little change in sensitivity to sulfonylurea and triazolopyrimidine herbicides [73, 74].

As with triazine resistance, double mutations have been identified that confer higher levels of resistance to ALS inhibitors [75, 76]. Imidazolinone-resistant corn and wheat lines were selected *in vitro* in cell cultures or following seed mutagenesis resistant to various classes of ALS inhibitor [77, 66]. The development of selective uses for these herbicides may result in added selection pressure for resistant weeds, emphasizing the need for careful herbicide management to maintain the long-term usefulness of these herbicides [78, 15].

### *3.1.6. Resistance to PSII-inhibiting herbicides*

Triazine and phenylurea herbicides do so by binding to the plastoquinone (PQ)-binding site on the D1 protein in the PS II reaction center of the photosynthetic electron transport chain. The D1 protein is coded by the *psb*A gene. PS II herbicide has two major consequences: (a) a shortage of reduced NADP`, which is required for CO2 fixation; and (b) the formation of free radicals which cause photooxidation of important several molecules such as chlorophylls and unsaturated lipids in the chloroplast. Triazine (simazine) resistance in weeds (*Senecio vulgaris*) was first identified in the late 1960s [24, 79, 80]. Since then, resistance to triazine herbicides has been reported in several weed species that many of them have developed in corn monocultures in the North America and Europe [81, 82]. Most s-triazine resistant biotypes show a high level of cross-resistance to other s-triazine herbicides, a lower level of resistance to as-triazinones, but no cross-resistance to phenylurea herbicides [83]. In almost all cases, a Ser264 to Gly mutation in the D1 protein is responsible for conferring resistance in weed biotypes [79]. QB may yet availability this site and transmits electrons to the cytochrome b6/f complex from the PS II reaction center, while the herbicide is absent. Ser264 to Gly mutation has no impact on the affinity of substituted urea herbicides and other PS II electron transport inhibitors, although it decreases the binding affinity of s-triazine and as-triazine herbicides to the D1 protein [39]. Biotypes containing this mutation exhibit a resistance factor of 1000 at the binding site on the D1 protein and 100 at the whole plant level [25, 84]. A resistant biotype of *Portulaca oleracea* has a high level of resistance to atrazine and to linuron, which through a Ser264 to Thr mutation which is the first reportage about D1 Ser264 to Thr mutation in higher plants selected under field conditions. Formerly, in tobacco and potato, this mutation had only been elected through tissue culture [85–87]. Both the Ser264 to Gly and Ser264 to Thr mutations reduce the efficiency of photosynthetic electron transport in the absence of herbicide [88, 89]. Resistance mutations can occur at positions other than Ser264, and mutations at Ser264 do not necessarily confer herbicide resistance. Molecular analysis has revealed that mutations at or close to positions Ser264, Phe265, Phe255, and His215 can affect the binding of PQ or herbicides and play an important part in the development of resistance [79, 11]. These results indicate that a mutation at Ser264 does not necessarily lead to resistance. Several mutations at positions other than Ser264 have been identified that confer resistance to triazine herbicides. Recently, a Val219 to Ile mutation has been identified in *Poa annua* populations resistant to metribuzin and diuron [90]. Val219 to Ile and Ala251 to Val or Thr mutations, without a change at Ser264, were suggested to be responsible for triazine resistance in various cell culture lines of *Chenopodium rubrum* [91]. In Ref. [92], Trebst has discussed amino acid changes between positions 211 and 275, including Phe211 to Ser, Gly256 to Asp, and Leu275 to Phe that confer herbicide resistance in various organisms. Some researchers reported a Ser268 to Pro mutation in soybean cell culture that confers a high level of resistance to both triazine and phenylurea herbicides [93]. Negative cross-resistance has been reported in some instances in which a triazine-resistant biotype is hypersensitive to phenylureas and other PS II-inhibiting herbicides [25, 94]. A *Chlamydomonas* mutant (Phe255 to Tyr) displayed negative cross-resistance to diuron and atrazine-resistant biotypes of *Amaranthus cruentus*, and *Amaranthus hybridus* showed negative cross-resistance to bentazon and pyridate [92, 95].

### *3.1.7. Resistance to auxin-type herbicides*

Auxin-type herbicides that mimic the endogenous auxin indole acetic acid (IAA) are among the oldest weed control products in use today. Nevertheless, the molecular binding site has not been recognized and the correct mechanism of action is not excellent realized, despite years of intense study. These compounds can motivate protein biosynthesis, and on the other hand, inhibit cell division and growth at low and higher concentrations, usually in the meristematic regions, respectively. The cell wall plasticity and nucleic acid metabolism primarily affect these compounds. The most broadleaf weeds well control through synthetic auxins [96]. Resistance to these herbicides is uncommon, considering their history of intensive use in cereal cropping systems [97].

### 3.2. Non-target-site herbicide resistance

Non-target-site-based resistance (NTSR) can confer unpredictable cross-resistance to herbicides. The non-target-site-based resistance mechanisms can interfere with herbicide penetration, translocation, and accumulation at the target site. NTSR is a part of the plant stress response. As such, NTSR is a dynamic process unrolling over time that involves "protectors" directly interfering with herbicide action, and also regulators controlling "protector" expression. NTSR is thus a quantitative trait. Infiltration of the herbicide into the plant and translocation to its site of action, reposition of the herbicide at its site of action, and binding of the herbicide to its target protein are three stages of herbicide action [8].

#### *3.2.1. Decreased herbicide infiltration and displacement*

Decrease in infiltration of herbicide has been reported in resistant weeds and crops for every main herbicide modes of action, that is, glyphosate [6, 98] acetolactate synthase, and acetyl CoA carboxylase inhibitors [99, 100]. This is owing to variations in the some physical and chemical attributes such as physical and chemical properties of the hull in resistant weeds and crops that bring a decline in the maintenance of the herbicides dilution on the foliage and/or a decrease in the influence of herbicides infiltration via the hull. Decrease in dislocation of herbicides has been investigated by other researchers [10, 51]. This phenomenon initiates some limitation in the move of the herbicide in the weeds or crops and in some cases the herbicide compartmenting. Decreased herbicide infiltration and displacement is a main mechanism of resistance to several herbicides, for example, paraquat and glyphosate. It has been made clear that depend on the weed or crop specie, or on the single weed or crop, non-target-site herbicide resistance toward paraquat inclusive the limited translocation through xylem, sequestration in the cell wall or in the vacuole and decreased uptake into the leaf cells [51, 101]. Limited translocations through the xylem and/or the phloem and/or quick sequestration to the vacuole are events, which occur through the non-target-site herbicide resistance toward glyphosate [6, 51, 102].

#### *3.2.2. Enhanced herbicide degradation*

Enhanced herbicide degradation is certainly the most studied aspect of NTSR. Herbicide degradation is a multistep process involving the coordinated action of several types of enzyme which have several stages. These stages involved the transformation of molecule of herbicide to some hydrophilic metabolite (stage I), the conjugation of hydrophilic metabolites into a plant acceptor molecule (such as a sugar) (stage II), and the exportation of the metabolite(s) into the vacuole and/or the cell walls after additional conjugation, cleaving, and/or oxidation stages [8, 103, 104].

#### *3.2.3. Conservation versus the parallel recompense of herbicide action*

This type of mechanism has been best studied in the case of *Alopecurus myosuroides* NTSR to ACCase repressors. Acetyl-CoA carboxylase (ACCase) inhibitors—herbicides, by interrupting biosynthesis of fatty acids, bring the extrication of active oxygen species that harm the ingredients of cells. Non-target-site-based resistance is mostly conferred through an increase in the expression of peroxidases that support the cells versus oxidative harm in several resistant plants. Hereon, non-target-site-based resistance is not case to an increase in degradation of herbicide in resistant species than sensitive species [8, 101, 105, 106].

## 4. Methods of identify resistance

### 4.1. Target-site resistance

The primary herbicide-resistant weeds were seriously examined in the 1980s–1990s. As a rule, resistance was given by means of TSR components controlled by prevailing alleles at a nuclear locus [7, 51, 107–109].

To date, nuclear monogenic control of TSR has been identified to herbicide groups A, B, K1, K2, E, and G (**Figure 1**), while legacy of TSR to triazine herbicides (C) is cytoplasmic. TSR is particularly across the board to herbicides in groups A, B, and C [7, 51, 108, 109]. Late advances demonstrate that atomic monogenic TSR is less basic than already suspected. Albeit most TSR cases will be surely given by overwhelming or semi-predominant alleles [7, 19], latent control of TSR has been accounted for imperviousness to herbicides in gathering K

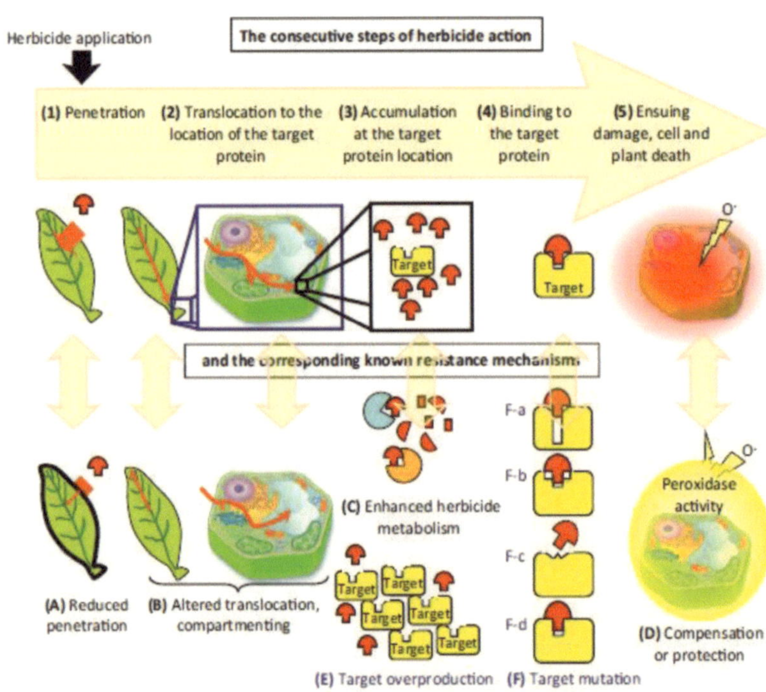


**Figure 1.** The action of herbicides following their application and the resistance mechanisms identified in weeds that correspond to each action step [122].

[110]. TSR is to a great extent invested by changes in the 3D structure of the herbicide target protein and in the dissemination of polar gatherings at positions significant for the security of herbicide official to the protein (**Figure 1**) [111, 112]. Auxiliary changes are for the most part because of amino-corrosive substitutions at one of a few conceivable positions on the herbicide target protein [7, 83, 108, 113]. A few substitutions giving resistance are conceivable at a given vital codon: upwards of 12 substitutions blessing resistance have been recognized at codon 197 in acetohydroxyacid synthase, the objective protein of gathering B herbicides [7]. The size of diminishment in proclivity of an herbicide for its coupling site depends both on the basic change in the objective protein and on the herbicide particle. Contingent upon the herbicide, a given basic change in the objective protein can give high or direct resistance [114, 115] or, in uncommon examples, an expansion in affectability to the herbicide (**Figure 1**) [110]. Along these lines, as opposed to the finishes of early reviews, the rising picture of TSR is not in highly contrasting, but rather in shades of dark [116].

It has been made clear that complex hereditary changes in weeds, including the erasure of a whole codon, progressive amino-corrosive substitutions coming about because of two sequential nucleotide substitutions at a similar codon, gathering of two amino-corrosive substitutions at particular codons that expanded the resistance level contrasted with a solitary transformation, and an expansion in amalgamation of the objective protein [117–119]. These systems seem from now on occasional in weeds, potentially because they include hereditary variations with a low likelihood of outward. In any case, the parallels with TSR to fungicides and insecticides propose that future work into the hereditary qualities of TSR to herbicides may uncover more perplexing components. Advancement of TSR is expected to adjust to the specific breadth model of adjustment [120] where a solitary valuable change of vast impact permits the underlying survival of mutants and after that spreads rapidly due to positive determination [121]. Basic populace hereditary models have demonstrated accommodating to coordinate the impacts of these developmental calculates the past and to evaluate the adequacy of different administration techniques in diminishing the likelihood of, and time to, resistance advancement [122].

Most DNA-based examinations for herbicide resistance depend on the polymerase chain reaction (PCR) to amplify a DNA sequence of interest from the milieu of DNA that is not of interest. Most standard "genomic" DNA extraction strategies yield DNA from the nuclear, chloroplastic, and mitochondrial genomes, and hence are appropriate for an extensive variety of downstream molecular analyses, including PCR. DNA can be removed from a wide range of plant material. In the absence of fresh tissue, high-quality DNA can also be extracted from preserved material [13].

DNA can as well as be synthesized from messenger RNA (mRNA) utilizing a reverse-transcriptase enzyme. This enzyme synthesizes DNA complementary to RNA (cDNA) from the 39 end of a primer hybridized on the RNA strand, utilizing the RNA strand as a template. cDNA is of specific interest when working on genes with complex intron–exon structure, because, like mRNAs, cDNAs do not contain introns [123]. The polymerase chain reaction (PCR) can hugely reproduce a given DNA district (amplicon) from little amounts of DNA. The easiest way to acquire sequence data for a given gene is to use the PCR amplicon as a template for

Sanger sequencing [13, 124]. In many studies, these approaches were used to uncover TSR in weeds [125–131].

### 4.2. Non-target-site resistance

Recognizing alleles of non-target-site-based resistance requires the identification of alleles specific for resistant genotypes while contrasted with sensitive genotypes, and to eliminate "false positives." These alleles are diverse in both genotypes (resistant and susceptible), however, do not involve in non-target-site-based resistance. In plant genomes, there are numerous alleles reported to be associated with non-target-site-based herbicide resistance. Alleles associated with quantitative characteristics are mostly identifies using genetic marker approaches (quantitative trait loci (QTL) mapping) [132]. QTL mapping is intricate, time consuming, and not easily applied to natural or field populations of nonmodel organisms such as weeds [133]. Another approach utilized for identification of alleles dedicating quantitative properties is to interrupt or imitate the phenotype of interest through genetic transformation [134]. Owing to the recent technical and scientific "omics" revolution, the genetic basis of quantitative characteristics, such as NTSR, can be explained even in nondemonstrate species, for example weeds. To accomplish this objective, three stages ought to be completed (**Figure 2**).

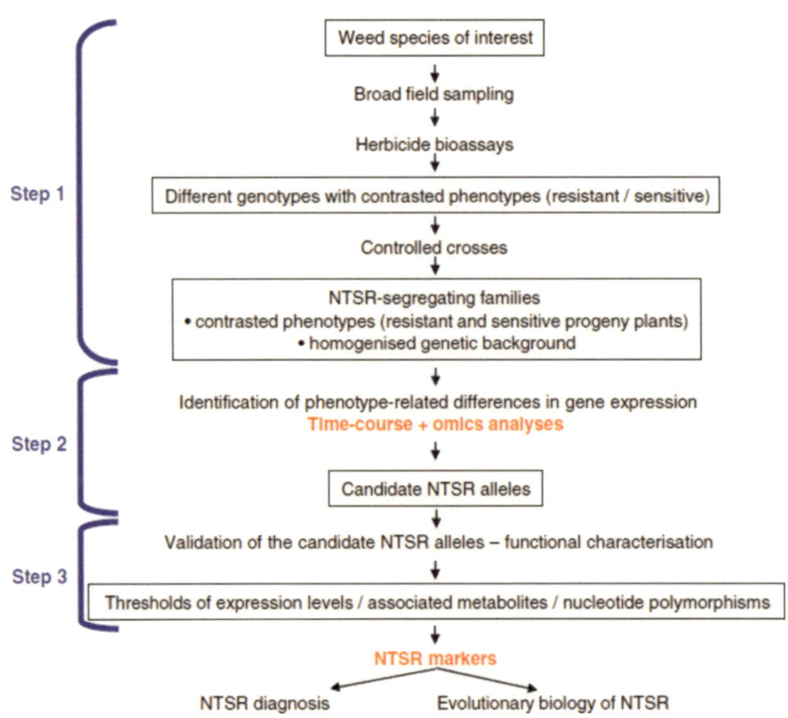


**Figure 2.** Three-step procedure to identify NTSR alleles [8].

http://dx.doi.org/10.5772/intechopen.69211

## 5. Cytochrome P450 monooxygenases (P450S) and evolved herbicide resistance

P450s are one of the largest super families of enzymes. Plants have the most noteworthy number of P450 genes. It is well demonstrated that tolerance to several modes of action to some herbicides is associated with Cyt P450 mediated enhanced metabolism. Cytochrome P450s in plants synthesize sterols, fatty acid derivatives, and hormones and in some cases, these are involved in plant secondary metabolism. Usually, plant cytochrome P450s are limited in the endoplasmic reticulum (in a few cases to plastid membranes). When a vast variety in plant metabolism catalyzes by cytochrome P450s, the role of cytochrome P450s is often hydroxylation or dealkylation in the herbicide alteration. Several plant cytochrome P450s will metabolize some herbicides to further inactivated productions so that their phytotoxicity will decrease or alter. Mostly, this process occurs by conjugation to glucose and subsequent transport into the vacuole. [36, 46, 135]. As crops can P450 metabolize many different herbicides, their use on large weed populations is a strong selection pressure for weed individuals possessing the same ability. Indeed, in weeds, P450-based herbicide resistance is a very threatening resistance mechanism because P450 enzymes can simultaneously metabolize herbicides of different modes of action, potentially including never-used herbicides [136, 137]. Subsequently, *in vivo* studies on herbicide metabolism and P450 inhibitors in resistant biotypes showed that P450s catalyzed enhanced rates of metabolism of several herbicides [138–140]. In addition to *L. rigidum* and *A. myosuroides*, the evolution of resistance due to P450-catalyzed enhanced rates of herbicide metabolism has been demonstrated in resistant biotypes of a further nine weed species [57, 141]. Cytochrome P450s can metabolize low dose of diclofop-methyl in some susceptible biotypes of *L. rigidum* and therefore the weeds were treated at a dose bringing around 50% mortality. Some survivors were grown for producing seeds to create the next generation, and in this direction, the selection was repeated. Non-target-site resistance was progress of high level in lonely three generations [142]. Importantly, there was concomitant evolution of cross-resistance to other P450-metabolizable herbicides of different modes of action [85, 88]. In *L. rigidum*, evolved P450-based herbicide resistance can be correlated with a fitness cost [143, 144]. Up to now, slight data are obtained through biochemical investigation to determine P450-based herbicide resistance in some evolved resistant weed species. Cytochrome P450 microsomes which degrade herbicides have not been successfully isolated from resistant *A. myosuroides* and *L. rigidum*, while they have been successfully isolated from resistant *E. phyllopogon* [145]. Isolated 16 P450 genes from a resistant *L. rigidum* biotype were not imputed to herbicide metabolism [76, 146]. One of three full-length P450s, CYP71R4, were gained from resistant *L. rigidum* biotypes metabolized a PSII herbicide, while expressed in yeast [147]. P450 proteins can share as little as 16% amino acid identity, and there are more than 2000 plant P450 sequences in the P450 database. Reportages about P450-based evolved herbicide resistance in grass weed species has often been more than dicot species, whereas dicot weed species have less P450 genes toward grass weed species. This fact reflects the attitude of some investigate to test only for target-site-based resistance [17].

## 6. Glutathione S-transferases and evolved herbicide resistance

Glutathione S-transferases (GSTs) which catalyze the conjugation of glutathione to variety of hydrophobic, electrophilic substrates, are multifunctional enzymes. Glutathione S-transferases (GSTs) have a special role in protecting the plant from oxidative stress (e.g., from reactive oxygen species), thus functioning as protective mechanism [16]. Glutathione S-transferases (GSTs) detoxify several herbicides in some crop and weed species. These enzymes play a role in stress response [148–150]. Glutathione-conjugated herbicides can be sequestered in the vacuole or exuded via root tips [149, 151]. Herbicide-metabolizing GSTs have been purified and characterized from several crops [150, 152]. Some studies such as molecular modeling, mutagenesis studies, and also the resolution of the 3D structure of plant GST (including herbicide-induced GST) provide an understanding of the molecular basis of GST-catalyzed herbicide binding and how single amino acid substitution(s) can improve GST catalytic efficiency and affect substrate specificity for herbicides and xenobiotics [153–155]. Because the Glutathione S-transferases (GSTs) catalyze the conjugation of triazines to glutathione through their high activity, these herbicides are selective for corn. This feature cause to widespread utilize of triazines can elect some weeds with glutathione S-transferases capable to eliminate them. Actually, in some weed species such as *Abutilon theophrasti*, developed GST-intervened triazine herbicide resistance has been observed [49, 156]. More researches demonstrated that enhanced activity of glutathione S-transferases is owing to higher catalytic susceptibility compared with overexpression enzyme or presence of a novel glutathione S-transferases [157]. This shows a conceivable transformation (mutation) in the gene of glutathione S-transferase gene which could better herbicide binding and so glutathione S-transferase catalytic performance. Resistance to atrazine as a singular nuclear gene with sectional predomination is inherited in this biotype [5]. It was demonstrated that in a resistant *Echinochloa phyllopogon* biotype, fenoxaprop-*p*-methyl resistance can be due to glutathione-herbicide conjugation [80]. Investigates with multiple resistant *A. myosuroides* biotypes with increased P450-catalyzed herbicide metabolism also show that they have higher GST activity [149, 158, 159]. Generally, GST enzymes can play both a direct role and an indirect role in evolved herbicide resistance [17].

## 7. Taxonomic effects in herbicide-resistant weeds and deployment of resistant crops

Evolved herbicide resistance (EHR) has become a threat to agriculture around the world [12, 160, 161]. Evolved herbicide resistance in weeds was initially reported in 1970 and generally considered during the 1970s throughout the 1990s [80, 162]. The rate of instances has precipitated significantly during those decades. Up to now, the advancement of imperviousness to various herbicides with various mode of action has additionally been detected inside various weed species [51]. The detection of resistance to glyphosate, and the introduction of transgenic glyphosate-resistant crops in the 1990s, also the recent expansion of cases of evolved resistance to glyphosate in weeds, likely because to greater glyphosate usage, have inspired a renewal of interest and resurgence of research into this phenomenon [113, 163].

In spite of four decades of research on evolved herbicide resistance, it is unclear wherefore a few weeds develop resistance quicker than others. Baker's list of specifications which may be anticipated in the "ideal weed" is excellent recognized; one may anticipate that weeds with evolved herbicide resistance will have a subset of these qualities [164]. This perception was ascribed to chance, as lots of resistant weeds among the world's worst weeds, are wide-spreading, and happen in many cropping systems [165–167]. Well before evolved herbicide resistance was detected, inheritable variability, breeding system, reproductive valence, and population size were predicted to associate with development of herbicide resistance [168]. Other plant variables can influence the development of resistance, including change recurrence, generation time, and compatibility in lack of the herbicide, pliancy, and soil seed repository, and in addition, method of legacy of resistance, size of population, seed dormancy, and gene flow by pollen and seed [121, 169]. Whenever these factors have been tested in models predicting evolution of resistance, few have been examined empirically [168, 170].

In spite of relatively cohesive internally of taxonomic families, there are usual differences in terms of ecological properties among them; in fact, evolved herbicide resistance does not occur randomly among weed or crop species. Generally, depending on perceptions and reportages of the tendency for resistance to evolve within certain genera or species, evolved herbicide-resistant weeds are distinct ecologically and taxonomically toward other weeds [171]. Some researchers found the same trends for subsets of weeds with EHR to acetolactate synthase (ALS), photosystem II (PSII), and 5-enolpyruvylshikimate-3-phosphate (EPSP) synthase-inhibitor herbicides and with multiple resistances. Comparing taxonomic and life history traits of weeds with EHR to a control group ("the world's worst weeds"), we found weeds with EHR significantly overrepresented in certain plant families and having certain life history biases [171].

## 8. GM crops

Herbicide-resistant crops (HRCs), sometimes called herbicide-tolerant crops, are crops made resistant to herbicides either by transgene technology or by selection in cell or tissue culture for mutations that confer resistance [172, 173]. Two techniques have been for all intents and purposes connected to generation of HR plants by hereditary building monetarily are: the presentation of the quality encoding the herbicide-inactivating catalyst, the presentation of the mutant, or outside quality. The advance in quality cloning and quality exchange, particularly utilizing *Agrobacterium tumefaciens*, has made hereditary building the most well known at present [174].

### 8.1. The gene encoding the herbicide-inactivating enzyme

This strategy has been most widely applied for the production of HR crops. The key step is to clone the gene encoding a herbicide-inactivating or detoxifying enzyme with high specificity and efficiency. Glufosinate-resistant crops sold as Liberty Link were produced by the introduction of the bar gene encoding the glufosinate-inactivating enzyme. The *bar* gene was cloned from Streptomyces hygroscopicus which produced bialaphos, the precursor of glufosinate (phosphinothricin). The *bar* gene encodes phosphinothricin-N-acetyl-transferase (PAT) which acetylates bialaphos to an inactivated form and prevents autotoxicity of bialaphos in the bacterium [174].

Bromoxynil-resistant crops sold as BXN were produced by the introduction of the *bxn* gene encoding the bromoxynil-inactivating enzyme [175]. However, the *gox* gene encoding the glyphosate-inactivating enzyme was also introduced into some plant species, possibly to enhance the level of resistance [176]. The gene encoding the 2,4-dichlorophenoxyacetic acid (2,4-D)-inactivating enzyme was also cloned from a soil bacterium because 2,4-D was readily inactivated in soil [177]. *Alcaligenes eutrophus,* thus selected, utilized 2,4-D as the sole source of carbon. The *tfd*A gene from the bacterium converts 2,4-dichlorophenoxyacetic acid (2,4-D) to 2,4-dichlorophenol. An option technique is to use the quality encoding the catalyst required in the imperviousness to more than one herbicide, for example, glutathione S-transferase and cytochrome P-450 [178].

Resistance types of EPSPS from petunia or Salmonella, conveying Gly96 to Ala or Pro101 to Ser changes, were assessed in early transgenic plants. In spite of the fact that these changes do give imperviousness to glyphosate, the reactant properties of the modified protein are hindered, which decreases the force of the plants without herbicide. The popularized glyphosate-safe harvests contain the *Agrobacterium* CP4 EPSPS quality [176]. Focusing on the CP4 quality to the chloroplast presents an abnormal state of imperviousness to glyphosate sans the negative impacts connected with the single transformation EPSPS qualities portrayed previously. In some transgenic edit cultivars, glyphosate resistance is presented by a mix of the CP4 quality and a bacterial oxidoreductase quality that detoxifies glyphosate. An option wellspring of glyphosate resistance for transgenic plants is an EPSPS quality conveying two separate changes in a similar district of the quality (positions 101, 102 as well as 106) [179]. This two-fold mutant quality has not been presented in any marketed safe yields. Glyphosate resistance can likewise be founded on intensification of the EPSPS quality, prompting to expanded levels of transcript creation and EPSPS action [180–182]. Essentially, adequate EPSPS is created to titrate out the glyphosate, leaving an overabundance of chemical that remaining parts practical. This has been appeared in plants chose in tissue culture or through repetitive choice. Now and again overexpression is lost when the choice weight is expelled or when plants are recovered; in others, the quality has all the earmarks of being steady in recovered plants and their descendants. Be that as it may, this component has not been utilized to create glyphosate-safe yields [15].

## 8.2. Mutant or foreign gene encoding the target enzyme with low affinity to the herbicide

This strategy is applicable to the production of any HR crops. But it was applied to commercial HR crops, which have been restricted to glyphosate-resistant crops sold as Roundup Ready. Generally, when an enzyme with a high herbicide binding constant was produced by a mutant gene, its enzymological characteristics were found to be unfavorable for the maximal enzyme activity leading to decreased growth and fitness of the plants transformed with this gene. It was known that EPSPSs from some bacteria were naturally resistant to glyphosate. EPSPS from *Agrobacterium* sp. strain CP4 was selected with high glyphosate-resistance and catalytic efficiency in the presence of glyphosate. The CP4 EPSPS gene was cloned from the bacterium and used for the production of Roundup Ready crops such as soybean, canola, cotton, maize, and sugar beet [51]. The target enzyme of sulfonylurea, imidazolinone, and triazolopyrimidine herbicides is acetolactate synthase (ALS). Various resistant ALS genes were cloned from tobacco [183] and *Arabidopsis thaliana* [184]. These gene products showed different levels of resistance to sulfonylureas, imidazolinones, and triazolopyrimidines [15]. These resistant ALS genes were introduced into plants individually, or in combination, and conferred resistance to these herbicides. Though some of these genes conferred resistance even at field trials, these genes have not been used for commercialization [185].

### 8.3. Novel tools for development of herbicide-resistant crops

Plant cells have three genomes and, in some plant seeds, two of these genomes are transformable: the nuclear genome and the genome of the plastids (chloroplasts). The plastid genome of photosynthetically active seed plants is a small circularly mapping genome of 120–220 kb, encoding 120–130 genes. It can be engineered by genetic transformation in a (still relatively small) number of plant species, and this possibility has stirred enormous interest among plant biotechnologists. There are considerable attractions associated with placing trans-genes into the plastid genome rather than the nuclear genome. First and foremost, the high number of plastids per cell and the high copy number of the plastid genome per plastid offer the possibility of expressing foreign genes to extraordinarily high levels, often one to two orders of magnitude higher than what is possible by expression from the nuclear genome [186, 187]. Second, transgene integration into the plastid genome occurs exclusively by homologous recombination, making plastid genome engineering a highly precise genetic engineering technique for plants. Third, as a prokaryotic system that is derived from a cyanobacterium acquired by endo-symbiosis, the plastid genetic system is devoid of gene silencing and other epigenetic mechanisms that interfere with stable transgene expression. Fourth, similar to bacterial genes, many plastid genes are arranged in operons offering the possibility to stack transgenes by arranging them in artificial operons. Finally, plastid transformation has received significant attention as a superb tool for transgene containment due to the maternal mode of plastid inheritance in most angiosperm species, which drastically reduces transgene transmission through pollen [188, 189]. Since the development of plastid transformation for the seed plant tobacco (*Nicotiana tabacum*) more than 20 years ago [64, 190], the community has assembled a large toolbox for plastid genetic engineering and also made some progress with developing plastid transformation protocols for additional species. Unfortunately, plastid transformation is still restricted to a relatively small number of species and not a single monocotyledonous species (including the cereals representing the world's most important staple foods) can be transformed. Thus, developing protocols for important crops continues to pose a formidable challenge in plastid biotechnology and significant strides forward are likely to require conscientious efforts and long-term investments in both the academic and the industrial sectors [26].

## 9. Conclusion

Herbicide-resistant weeds are a crucial topic in agriculture. Growers need to interchange weed management techniques thus prolonging the development of herbicide-resistance in weeds. Defiantly, the important and effective approaches to manage the herbicide-resistant weeds are prevention of weed emergence, integrated application of all available options to weed control, and rotate herbicides with different modes of action.

Nowadays, Herbicide-resistant crops have transformed the weed management strategy of many growers. Herbicide-resistant or - tolerant crops have helped farmers to manage weeds more comfortable to meet the growing demands for human food, fiber, and fuel and animal feed. The advent and development of herbicide-resistant crops has provided conditions to minimize production losses because of weed infestation. Generally, growers need intelligent management approaches to maximize the long-term benefits of this technology and reduce weed shifts to difficult-to-control and herbicide-resistant weeds.

## Author details

Mohammad Taghi Alebrahim[1*], Rouzbeh Zangoueinejad[1] and Te Ming Tseng[2]

*Address all correspondence to: m_ebrahim@uma.ac.ir

1 Department of Agronomy and Plant Breeding, Faculty of Agriculture and Natural Resources, University of Mohaghegh Ardabili, Ardabil, Iran

2 Department of Plant and Soil Sciences, Mississippi State University, Mississippi State, MS, USA.

## References

[1] Freed VH. Weed science: The emergence of a vital technology. Weed Sciences. 1980;**28**:621

[2] Avery DT. Saving the Planet with Pesticides and Plastic. Indianapolis: Hudson Institute; 1992

[3] Barry G, Kishore G, Padgette M, Kolacz K, Weldon M, Re D, Eichholtz D, Fincher K, Hallas L. Inhibitors of amino acid biosynthesis: Strategies for imparting glyphosate tolerance to crop plants. In: Singh BK, Flores HE, Shannon JC, editors. Biosynthesis and Molecular Regulation of Amino Acids in Plants. Rockville, MD: American Society of Plant Physiologists; 1992. pp. 139-145

[4] Herms DA, Mattson WJ. The dilemma of plants, to grow or to defend. The Quarterly Review of Biology. 1995;**67**:283

[5] Powles SB, Holtum JAM, editors. Herbicide Resistance in Plants: Biology and Biochemistry. Boca Raton, FL: Lewis Publishers; 1994. p. 353

[6] Vila-Aiub MM, Balbi MC, Distéfano AJ, Fernández L, Hopp E, Yu Q, et al. Glyphosate resistance in perennial *Sorghum halepense* (Johnsongrass), endowed by reduced glyphosate translocation and leaf uptake. Pest Management Science. 2012;**68**:430-436

[7] Beckie H, Tardif FJ. Herbicide cross resistance in weeds. Crop Protection. 2012;**35**:15-28

[8] Délye C. Unravelling the genetic bases of non-target-site based resistance (NTSR) to herbicides: A major challenge for weed science in the forthcoming decade. Pest Management Science. 2012;**69**:176-187

[9] Yih RY, McRae H, Wilson HF. Mechanism of selective action of 3,4-dichloropropionanilide. Plant Physiology. 1968;**4**:1291

[10] Zagnitko O, Jelenska J, Tevzadze G, Haselkorn R, Gornicki P. An isoleucine/leucine residue in the carboxyl transferase domain of acetyl-CoA carboxylase is critical for interaction with aryloxyphenoxypropionate and cyclohexanedione inhibitors. Proceedings of the National Academy of Sciences of the United States of America. 2001;**98**:6617-6622

[11] Devine MD, Eberlein CV. Physiological, biochemical and molecular aspects of herbicide resistance based on altered target sites. In: Roe RM, Burton JD, Kuhr RJ, editors. Herbicide Activity: Toxicology, Biochemistry and Molecular Biology. Amsterdam: IOS Press; 1997. pp. 159-185

[12] Yuan JS, Tranel PJ, Stewart Jr. CN. Non-target-site herbicide resistance: A family business. Trends in Plant Science. 2007;**12**:6-13

[13] Délye C, Duhoux A, Pernin F, Riggins CW, Tranel PJ. Molecular mechanisms of herbicide resistance. Weed Sci. 2015;**63**(Special Issue):91-115

[14] Yang Q, Deng W, Li X, Yu Q, Bai L, Zheng M. Target-site and non-target-site based resistance to the herbicide tribenuron-methyl in flixweed (*Descurainia sophia* L.). BMC Genomics. 2016;**17**:551

[15] Devine MD, Shukla A. Altered target sites as a mechanism of herbicide resistance. Crop Protection. 2000;**19**:881-889

[16] Powles SB, Preston C, Bryan IB, Jutsum AR. Herbicide resistance: Impact and management. Advances in Agronomy. 1997;**58**:57

[17] Preston C, Mallory-Smith CA. Biochemical mechanisms, inheritance, and molecular genetics of herbicide resistance in weeds. In: Powles SB, Shaner DL, editors. Herbicide Resistance and World Grains. USA: CRC Press LLC; 2001. pp. 34-71

[18] Ichinose K, Che FS, Kimura Y, Matsunobu A, Sato F, Yoshida Y. Selection and characterization of protoporphyrinogen oxidase inhibiting herbicide (S23412) resistant photomixotrophic cultured cells of Nicotiana tabacum. Journal of Plant Physiology. 1995;**146**:693-698

[19] Randolph-Anderson BL, Sato R, Johnson AM, Harris EH, Rasche E. Glufosinate ammonium tolerant crops - international commercial developments and experiences. Brighton Crop Protection Conference Weeds. 1997. pp. 941-946

[20] Duke SO, Lee HJ, Duke MV, Reddy KN, Sherman TD, Becerril JM, et al. Mechanisms of resistance to protoporphyrinogen oxidase-inhibiting herbicides. In: De Prado R, Jorrin J, Garcia-Torres L, editors. Weed and Crop Resistance to Herbicides. Dordrecht, Netherlands: Kluwer Academic Publishers; 1997. pp. 155-160

[21] Hall LM, Moss SR, Powles SB. Mechanism of resistance to chlorotoluron in two biotypes of the grass weed *Alopecurus myosuroides*. Pesticide Biochemistry and Physiology. 1995;**53**:180

[22] Matsunaka S, Ito K. Paraquat resistance in Japan. In: Caseley JC, Cussans GW, Atkin RK, editors. Herbicide Resistance in Weeds and Crops. Oxford: Butterworth- Heinemann; 1991. p. 77

[23] Heap I. International Survey of Herbicide Resistant Weeds [Internet]. 2000. Available from: www.weedscience.com

[24] Devine MD, Duke SO, Fedtke C. Physiology of Herbicide Action. Englewood Cliffs, NJ: Prentice-Hall; 1993. p. 441

[25] Fuerst EP, Arntzen CJ, Pfister K, Penner D. Herbicide cross-resistance in triazine-resistant biotypes of four species. Weed Science. 1986;**34**:344-353

[26] Bock R. Genetic engineering of the chloroplast: Novel tools and new applications. Current Opinion in Biotechnology. 2014;**26**:7-13

[27] Gressel J. Why get resistance? It can be prevented or delayed. In: Caseley JC, Cussans GW, Atkin RK, editors. Herbicide Resistance in Weeds and Crops. Oxford, UK: Butterworth-Heinemann; 1991. pp. 1-26

[28] Anthony RG, Waldin TR, Ray JA, Bright SWJ, Hussey PJ. Herbicide resistance caused by spontaneous mutation of the cytoskeletal protein tubulin. Nature. 1998;**393**:260-263

[29] Yamamoto E, Zeng L, Baird WV. A-tubulin missense mutations correlate with antimicrotubule drug resistance in *Eleusine indica*. Plant Cell. 1998;**10**:297-308

[30] Anthony RG, Reichelt S, Hussey PJ. Dinitroaniline herbicide- resistant transgenic tobacco plants generated by co-expression of a mutant a-tubulin and a b-tubulin. Nature Biotechnology. 1999;**17**:712-716

[31] Lee VD, Huang B. Missense mutations at lysine 350 in b-tubulin confer altered sensitivity to microtubule inhibitors in Chlamydomonas. Plant Cell. 1990;**2**:1051-1057

[32] Boyko A, Kovalchuk I. Epigenetic control of plant stress response. Environmental and Molecular Mutagenesis. 2008;**49**:61-72

[33] Carey VF, Hoagland RE, Talbert RE. Verification and distribution of propanil-resistant barnyardgrass in Arkansas. Weed Technology. 1995;**9**:366

[34] Mendenez J, De Prado R. Diclofop-methyl cross-resistance in a chlorotoluronresistant biotype of Alopecurus myosuroides. Pesticide Biochemistry and Physiology. 1996;**56**:123

[35] Incledon BJ, Hall JC. Acetyl-coenzyme A carboxylase: Quanternary structure and inhibition by graminicidal herbicides. Pesticide Biochemistry and Physiology. 1997;**57**:255

[36] Sabba RP, Sterling TM, Lownds NK. Effect of picloram on resistant and susceptible yellow starthistle: The role of ethylene. Weed Science. 1998;**46**:297

[37] Singh S, Kirkwood RC, Marshall G. Effect of ABT on the activity and rate of degradation of isoproturon in susceptible and resistant biotypes of *Phalaris minor* and in wheat. Journal of Pesticide Science. 1998;**53**:123

[38] Lee LJ, Ngim J. A first report of glyphosate-resistant goosegrass (*Eleusine indica* (L) Gaertn) in Malaysia. Pest Management Science. 2000;**56**:336

[39] Fernandez P, Gauvrit C, Barro F, Menendez J, De Prado R. First case of glyphosate resistance in France. Agronomy for Sustainable Development. 2015

[40] Gornicki P, Faris J, King I, Podowinski J, Gill B, Haselkorn R. Plastidlocalized acetyl-CoA carboxylase of bread wheat is encoded by a single gene on each of the three ancestral chromosome sets. Proceedings of the National Academy of Sciences of the United States of America. 1997;**94**:14179

[41] Islam AKMR, Powles SB. Inheritance of resistance to paraquat in barley grass *Hordeum glaucum* Steud. Weed Research. 1998;**28**:393

[42] Guttieri MJ, Eberlein CV, Mallory-Smith CA, Thill DC. Molecular genetics of target-site resistance to acetolactate synthase inhibiting herbicides. In: Brown TM, editor. Molecular Genetics and Evolution of Pesticide Resistance, ACS Symp. Ser. 645. Washington, DC: American Chemical Society; 1996. p. 10

[43] Wrubel RP, Gressel J. Are herbicide mixtures useful for delaying the rapid evolution of resistance? A case study. Weed Technology. 1994;**8**:635

[44] Eberlein CV, Guttieri MJ, Berger PH, Fellman JK, Mallory-Smith CA, Ernst D, Kiefer E, Bühler M, Bogenrieder A, Sandermann Jr. H. On the basis of atrazine resistance in weeds. In: De Prado R, Jorrin J, Torres LG, Marshall G, editors. Proceedings of the International Symposium on Weed and Crop Resistance to Herbicides. Cordoba, Spain: Graficas TYPO; 1996. pp. 56-59

[45] Devine MD. Mechanisms of resistance to acetyl-coenzyme A carboxylase inhibitors: A review. Journal of Pesticide Science. 1997;**51**:259

[46] Shaaltiel Y, Glazer A, Bocion TF, Gressel J. Cross-tolerance to herbicides and environmental oxidants of plant biotypes tolerant to paraquat, sulfur dioxide, and ozone. Pesticide Biochemistry and Physiology. 1988;**31**:12

[47] Ribeiro DN, Pan Z, Duke SO, Nandula VK, Baldwin BS, Shaw DR, Dayan FE. Involvement of facultative apomixis in inheritance of EPSPS gene amplification in glyphosate-resistant *Amaranthus palmeri*. Planta. 2013;**239**:199-212

[48] Fernández P, Alcántara R, Osuna MD, MartinM Vila-Aiub MM, De Pradoa R. Forward selection for multiple resistance across the non-selective glyphosate, glufosinate and oxyfluorfen herbicides in *Lolium* weed species. Pest Management Science. 2016

[49] Burton JD, Gronwald JW, Somers DA, Connelly JA, Gengenbach BG, Wyse DL. Inhibition of plant acetyl-CoA carboxylase by the herbicides sethoxydim and haloxyfop. Biochemical and Biophysical Research Communications. 1987;**148**:1039-1044

[50] Rendina AR, Felts JM. Cyclohexanedione herbicides are selective and potent inhibitors of acetyl-CoA carboxylase from grasses. Plant Physiology. 1988; **86**:983-986

[51] Powles SB, Yu Q. Evolution in action: Plants resistant to herbicides. Annual Review of Plant Biology. 2010;**61**:317-347

[52] Konishi T, Sasaki Y. Compartmentalization of two forms of acetyl-CoA carboxylase in plants and the origin of their tolerance towards herbicides. Proceedings of the National Academy of Sciences of the United States of America. 1994;**91**:3598-3601

[53] Devine MD, Shimabukuro RH. Resistance to Acetyl coenzyme A Carboxylase Inhibiting Herbicides. In: Powles SB, Holtum JAM, editors. Herbicide Resistance in Plants: Biology and Biochemistry. Boca Raton, FL: Lewis Publishers; 1994. pp. 141-169

[54] Catanzaro CJ, Burton JD, Skroch WA. Graminicide resistance of acetyl-CoA carboxylase from ornamental grasses. Pesticide Biochemistry and Physiology. 1993;**45**:147-153

[55] Stoltenberg DE, Gronwald JW, Wyse DL, Burton JD, Somers DA, Gengenbach BG. Effect of sethoxydim and haloxyfop on acetyl-coenzyme A carboxylase activity in Festuca species. Weed Science. 1989;**37**:512-516

[56] Marles MAS, Devine MD, Hall JC. Herbicide resistance in *Setaria viridis* conferred by a less sensitive form of acetyl coenzyme A carboxylase. Pesticide Biochemistry and Physiology. 1993;**46**:7-14

[57] Shukla A, Dupont S, Devine MD. Resistance to ACCase inhibitor herbicides in wild oat: Evidence for target site-based resistance in two biotypes from Canada. Pesticide Biochemistry and Physiology. 1997a;**57**:147-155

[58] Norman MA, Smeda RJ, Vaughn KC, Fuerst E P. Differential movement of paraquat in resistant and sensitive biotypes of Conyza. Pesticide Biochemistry and Physiology. 1994;**50**:31

[59] Leach GE, Devine MD, Kirkwood RC, Marshall G. Target enzyme-based resistance to acetyl-coenzyme A carboxylase inhibitors in *Eleusine indica*. Pesticide Biochemistry and Physiology. 1995;**51**:129-136

[60] Tardif FJ, Powles SB. Herbicide multiple-resistance in a *Lolium rigidum* biotype is endowed by multiple mechanisms: Isolation of a subset with resistant acetyl-CoA carboxylase. Physiology Plant. 1994;**91**:488-494

[61] Preston C, Tardif FJ, Christopher JT, Powles SB. Multiple resistance to dissimilar herbicide chemistries in a biotype of *Lolium rigidum* due to enhanced activity of several herbicide degrading enzymes. Pesticide Biochemistry and Physiology. 1996;**54**:123-134

[62] Gronwald JW, Eberlein CV, Betts KJ, Baerg RJ, Ehlke NJ, Wyse DL. Mechanism of diclofop resistance in an Italian ryegrass (*Lolium multiyorum* Lam.) biotype. Pesticide Biochemistry and Physiology. 1992;**44**:126-139

[63] Shukla A, Leach GE, Devine MD. High-level resistance to sethoxydim conferred by an alteration in target enzyme, acetyl-CoA carboxylase, in *Setaria faberi* and *Setaria viridis*. Plant Physiology and Biochemistry. 1997b;**35**:803-807

[64] Bourgeois L, Kenkel NC, Morrison IN. Characterization of cross-resistance patterns in acetyl-CoA carboxylase inhibitor resistant wild oat (*Avena fatua*). Weed Science. 1997;**45**:750-755

[65] Evenson KJ, Gronwald JW, Wyse DL. Isoforms of acetyl- Coenzyme A carboxylase in *Lolium multiyorum*. Plant Physiology and Biochemistry. 1997;**35**:265-272

[66] Nikolskaya T, Zagnitko O, Tevzadze G, Haselkorn R, Gornicki P. Herbicide sensitivity determinant of wheat plastid acetyl-CoA carboxylase is located in a 400-amino acid fragment of the carboxyltransferase domain. Proceedings of the National Academy of Sciences of the United States of America. 1999;**96**:14647-14651

[67] Zhang XQ, Devine MD. A possible mutation of plastidic ACCase gene conferring resistance to sethoxydim in green foxtail (*Setaria viridis*). Weed Science Society of America Abstracts. 2000;**40**:33

[68] Murray BG, Morrison IN, Brûlé-Babel A. Inheritance of acetyl-CoA carboxylase inhibitor resistance in wild oat (*Avena fatua*). Weed Science. 1995;**43**:233

[69] Norman MA, Fuerst EP, Smeda RJ, Vaughn KC. Evaluation of paraquat resistance mechanisms in Conyza. Pesticide Biochemistry and Physiology. 1993;**46**:236

[70] Vaughn KC, Fuerst EP. Structural and physiological studies of paraquat resistant Conyza. Pesticide Biochemistry and Physiology. 1985;**24**:86

[71] Anderson MP, Gronwald JW. Atrazine resistance in a velvetleaf (*Abutilon theophrasti*) biotype due to enhanced glutathione-S-transferase activity. Plant Physiology. 1991;**96**:104

[72] Saari LL, Cotterman JC, Thill DC. Resistance to acetolactate synthase-inhibitor herbicides. In: Powles SB, Holtum JAM, ediotrs. Herbicide Resistance in Plants: Biology and Biochemistry. Boca Raton, FL: Lewis Publishers; 1994. pp. 83-139

[73] Bernasconi P, Woodworth AR, Rosen BA, Subramanian MV, Siehl DL. A naturally occurring point mutation confers broad range tolerance to herbicides that target acetolactate synthase. Journal of Biological Chemistry. 1995;**270**:17381-17385

[74] Sathasivan K, Haughn GW, Murai N. Molecular basis of imidazolinone herbicide resistance in Arabidopsis thaliana var Columbia. Plant Physiology. 1991;**97**:1044-1050

[75] Creason GL, Chaleff RS. A second mutation enhances resistance of a tobacco mutant to sulfonylurea herbicides. Theoretical and Applied Genetics. 1988;**76**:177-182

[76] Wright TR, Bascomb NF, Sturner SF, Penner D. Biochemical mechanism and molecular basis for ALS-inhibiting herbicide resistance in sugarbeet (*Beta vulgaris*) somatic cell selections. Weed Science. 1998;**46**:13-23

[77] Newhouse KE, Smith WA, Starrett MA, Schaefer TJ, Singh BK. Tolerance to imidazolinone herbicides in wheat. Plant Physiology. 1992;**100**:882-886

[78] Boutsalis P, Karotam J, Powles SB. Molecular basis of resistance to acetolactate synthase-inhibiting herbicides in *Sisymbrium orientale* and *Brassica tournefortii*. Journal of Pesticide Science. 1999;**55**:507

[79] Gronwald JW. Resistance to photosystem II inhibiting herbicides. In: Powles SB, Holtum JAM, editors. Herbicide Resistance in Plants: Biology and Biochemistry. Boca Raton, FL: CRC Press; 1994. pp. 27-60

[80] Ryan GF. Resistance of common groundsel to simazine and atrazine. Weed Science. 1970;**18**:614-616

[81] LeBaron HM. Distribution and seriousness of herbicide-resistant weed infestations worldwide. In: Caseley JC, Cussans GW, Atkin RK, editors. Herbicide Resistance in Weeds and Crops. Oxford, UK: Butterworth- Heinemann; 1991. pp. 27-43

[82] LeBaron HM, McFarland J. Herbicide resistance in weeds and crops: An overview and prognosis. In: Green MB, LeBaron HM, Moberg WK, editors. Managing Resistance to Agrochemicals: From Fundamental Research to Practical Strategies. Washington, DC: American Chemical Society Symposium Series No. 421; 1990. pp. 336-352

[83] Shukla A, Devine MD, Mechanisms of selectivity and resistance to triazine herbicides. In: LeBaron HM, Gianessi LP, McFarland J, editors. The Triazine Herbicides. Boca Raton, FL: Lewis Publishers Inc.; 2000. in press

[84] Pfister K, Arntzen CJ. The mode of action of photosystem II-specific inhibitors in herbicide-resistant weed biotypes. Zeitschrift für Naturforschung. 1979;**34c**:996-1009

[85] Masabni JG, Zandstra BH. A serine-to-threonine mutation in linuron-resistant *Portulaca oleracea*. Weed Science. 1999;**47**:393-400

[86] Sigematsu Y, Sato F, Yamada Y. The mechanism of herbicide resistance in tobacco cells with a new mutation in the QB protein. Plant Physiology. 1989;**89**:986-992

[87] Smeda RJ, Hasegawa PM, Goldsbrough PB, Singh NK, Weller SC. A serine-to-threonine substitution in the triazine herbicide-binding protein in potato cells results in atrazine resistance without impairing productivity. Plant Physiology. 1993;**103**:911-917

[88] Masabni JG, Zandstra BH. Discovery of a common purslane (*Portulaca oleracea*) biotype resistant to linuron. Weed Technology. 1999;**13**:599-605

[89] McCloskey WB, Holt JS. Triazine resistance in *Senecio vulgaris* parental and nearly iso-nuclear backcrossed biotypes is correlated with reduced productivity. Plant Physiology. 1990;**92**:954-962

[90] Mengistu LW, Mueller-Warrant GW, Barker RE. Genetic diversity of Poa *annua* in western Oregon grass seed crops. Theoretical and Applied Genetics. 2000;**101**:70-79

[91] Schwenger-Erger C, Thiemann J, Barz W, Johanningmeier U, Naber D. Metribuzin resistance in photoautotrophic *Chenopodium rubrum* cell cultures. FEBS Letters. 1993;**329**:43-46

[92] Trebst A. The molecular basis of resistance of photosystem II herbicides. In: Caseley JC, Cussans GW, Atkin RK, editors. Herbicide Resistance in Weeds and Crops. Oxford, UK: Butterworth- Heinemann; 1991. pp. 145-164

[93] Alfonso M, Pueyo JJ, Gaddour K, Etienne A-L, Kirilovsky D, Picorel R. Induced new mutation of D1 serine-268 in soybean photosynthetic cell cultures produced atrazine resistance, increased stability of S 2 Q~B and S 3 Q~B states, and increased sensitivity to light stress. Plant Physiology. 1996;**112**:1499-1508

[94] Van Oorschot JL, Van Leeuwen PH. Inhibition of photosynthesis in intact plants of biotypes resistant or susceptible to atrazine and cross-resistant to other herbicides. Weed Research. 1988;**28**:223-230

[95] De Prado R, Sanchez M, Jorrin J, Dominiguez C. Negetive cross-resistance to betazone and pyridate in atrazine-resistant *Amaranthus cruentus* and *Amaranthus hybridus* biotypes. Journal of Pesticide Science. 1992;**35**:131-136

[96] Sterling TM, Hall JC. Mechanism of action of natural auxins and the auxinic herbicides. In: Roe RM, Burton JD, Kuhr RJ, editors. Herbicide Activity: Toxicology, Biochemistry and Molecular Biology. Amsterdam: IOS Press; 1997. pp. 111-141

[97] Webb SR, Hall JC. Auxinic-herbicide resistant and susceptible wild mustard (*Sinapis arvensis* L.) biotypes: Effect of auxinic herbicides on seedling growth and auxin-binding activity. Pesticide Biochemistry and Physiology. 1995;**52**:137-148

[98] Michitte P, De Prado R, Espinoza N, Ruiz-Santaella JP, Gauvrit C. Mechanisms of resistance to glyphosate in a ryegrass (*Lolium multiflorum*) biotype from Chile. Weed Science. 2007;**55**:435-440

[99] De Prado JL, Osuna MD, Heredia A, De Prado R. *Lolium rigidum*, a pool of resistance mechanisms to ACCase inhibitor herbicides. Journal of Agricultural and Food Chemistry. 2005;**53**:2185-2191

[100] White AD, Owen MDK, Hartzler MGand Cardina J. Common sunflower resistance to acetolactate synthase-inhibiting herbicides. Weed Science. 2002;**50**:432-437

[101] Preston C. Resistance to photosystem I disrupting herbicides. In: Powles SB, Holtum JAM, editors. Herbicide Resistance in Plants. Boca Raton, FL: Lewis Publishers; 1994. pp. 61-82

[102] Ge X, d'Avignon DA, Ackerman JJH, Sammons RD. Rapid vacuolar sequestration: The horseweed glyphosate resistance mechanism. Pest Management Science. 2010;**66**: 345-348

[103] Kreuz K, Tommasini R, Martinoia E. Old enzymes for a new job. Herbicide detoxification in plants. Plant Physiology. 1996;**111**:349-353

[104] Van Eerd LL, Hoagland RE, Zablotowicz RM, Hall JC. Pesticide metabolism in plants and microorganisms. Weed Science. 2003;**51**:472-495

[105] Cummins I, Bryant DN, Edwards E. Safener responsiveness and multiple herbicide resistance in the weed black-grass (*Alopecurus myosuroides*). Plant Biotechnology Journal. 2009;**7**:807-820

[106] Cummins I, Cole DJ, Edwards R. A role for glutathione transferases functioning as glutathione peroxidases in resistance to multiple herbicides in black-grass. Plant Journal. 1999;**18**:285-292

[107] Oerke E-C. Crop losses to pests. Journal of Agricultural Science. 2006;**144**:31-43

[108] Avila-Garcia WV, Sanchez-Olguin E, Hulting AG, Mallory-Smith C. Target-site mutation associated with glufosinate resistance in Italian ryegrass (*Lolium perenne* L. ssp. multiflorum). Pest Management Science. 2012;**68**:1248-1254

[109] Shaner DL, Lindenmeyer RB, Ostlie MH. What have the mechanisms of resistance to glyphosate taught us? Pest Management Science. 2012;**68**:3-9

[110] Délye C, Menchari Y, Michel S, Darmency H. Molecular bases for sensitivity to tubulin-binding herbicides in green foxtail. Plant Physiology. 2004;**136**:3920-3932

[111] Wang JG, Lee PK, Dong YH, Pang SS, Duggleby R G, Li ZM, Guddat LW. Crystal structures of two novel sulfonylurea herbicides in complex with Arabidopsis thaliana acetohydroxyacid synthase. FEBS Journal. 2009;**276**:1282-1290

[112] Yu LPC, Kim YS, Tong L. Mechanism for the inhibition of the carboxyltransferase domain of acetyl-coenzyme A carboxylase by pinoxaden. Proceedings of the National Academy of Sciences of the United States of America. 2010;**107**:22072-22077

[113] Powles SB, Lorraine-Colwill DF, Dellow JJ, Preston C. Evolved resistance to glyphosate in rigid ryegrass (*Lolium rigidum*) in Australia. Weed Science. 1998;**46**:604-607

[114] Délye C, Pernin F, Michel S. 'Universal' PCR assays detecting mutations in acetyl-coenzyme A carboxylase or acetolactate synthase that endow herbicide resistance in grass weeds. Weed Research. 2005;**51**:353-362

[115] Yu Q, Collavo A, Zheng M-Q, Owen M, Sattin M, Powles SB. Diversity of acetyl-coenzyme A carboxylase mutations in resistant Lolium populations: Evaluation using Bclethodim. Plant Physiology. 2007;**145**:547-558

[116] Leroux P, Walker AS. Multiple mechanisms account for resistance to sterol 14a demethylation inhibitors in field isolates of Mycosphaerella graminicola. Pest Management Science. 2011;**67**:44-59

[117] Han H, Yu Q, Edison Purba E, Li M, Walsh M, Friesen S, Powles SB. A novel amino acid substitution Ala-122-Tyr in ALS confers high-level and broad resistance across ALS inhibiting herbicides. Pest Management Science. 2012;**68**:1164-1170

[118] Jalaludin A, et al. Evolution in action: A double amino acid substitution in the EPSPS gene endows high-level glyphosate resistance. In: Powles S, Mayer L, editors. Global Herbicide Resistance Challenge. Australian Herbicide Resistance Initiative; 2013. p. 39

[119] Patzoldt WL, Hager AG, McCormick JS, Tranel PJ. A codon deletion confers resistance to herbicides inhibiting protoporphyrinogen oxidase. Proceedings of the National Academy of Sciences of the United States of America. 2006;**103**:12329-12334

[120] Pritchard JK, Pickrell JK, Coop G. The genetics of human adaptation: Hard sweeps, soft sweeps, and polygenic adaptation. Current Biology. 2010;**20**:R208–R215

[121] Jasieniuk MJ, Brû lé-Babel AL, Morrison IN. The evolution and genetics of herbicide resistance in weeds. Weed Science. 1996;**44**:176-193

[122] Délye1 C, Jasieniuk M, Corre VL. Deciphering the evolution of herbicide resistance in weeds. Trends in Genetics. 2013;**29(11)**:649-658

[123] Délye C, Zhang X-Q, Michel S, Matéjicek A, Powles SB. Molecular bases for sensitivity to acetyl-coenzyme A carboxylase inhibitors in black-grass. Plant Physiology. 2005;**137**:794-806

[124] Sanger F, Air GM, Barrell BG, Brown NL, Coulson AR, Fiddes CA, Hutchison CA, Slocombe PM, Smith M. Nucleotide sequence of bacteriophage phi X174 DNA. Nature. 1977;**265**(5596):687-695

[125] Christopher JT, Powles S, Holtum JAM. Resistance to acetolactate synthase-inhibiting herbicides in annual ryegrass (*Lolium rigidum*) involves at least two mechanisms. Plant Physiology. 1992;**100**:1909-1913

[126] Yu Q, Han H, Powles SB. Mutations of the ALS gene endowing resistance to ALS-inhibiting herbicides in *Lolium rigidum* populations. Pest Management Science. 2008;**64**:1229-1236

[127] Mohamed IA, Li R, You Z, Li Z. Japanese Foxtail (*Alopecurus japonicus*) Resistance to Fenoxaprop and Pinoxaden in China. Weed Science. 2012;**60**:167-171

[128] Doyle JJ, Doyle JL. A rapid DNA isolation procedure for small quantities of fresh leaf tissue. Phytochem. Bull. 1987;**19**:11-15

[129] Liu W, Harrison DK, Chalupska, D, Gornicki P, O'Donnell CC, Llewellyn D, Last D. Genetic engineering of crops for tolerance to 2,4-D. In: Duke SO, editor. Herbicide-resistant Crops. Boca Raton: Lewis; 1996. pp. 159-174

[130] Petit C, Bay G, Pernin F, Délye C. Prevalence of cross- or multiple resistance to the acetyl-coenzyme A carboxylase inhibitors fenoxaprop, clodinafop and pinoxaden in black-grass (Alopecurus myosuroides Huds.) in France. Pest Management Science. 2010;**66**:168-177

[131] Scarabel L, Panozzo S, Varottob S, Sattina M. Allelic variation of the ACCase gene and response to ACCase-inhibiting herbicides in pinoxaden-resistant *Lolium* spp. Pest Management Science. 2011;**67**:932-941

[132] Collard BCY, Jahufer MZZ, Brouwer JB, Pang ECK. An introduction to markers, quantitative trait loci (QTL) mapping and marker assisted selection for crop improvement: The basic concepts. Euphytica. 2005;**142**:169-196

[133] Slate J. Quantitative trait locus mapping in natural populations: Progress, caveats and future directions. Molecular Ecology. 2005;**14**:363-379

[134] Østergaard L, Yanofsk MF. Establishing gene function by mutagenesis in *Arabidopsis thaliana*. Plant Journal. 2004;**39**:682-696

[135] Yamasue Y, Kamiyama K, Hanioka Y, Kusanagi T. Paraquat resistance and its inheritance in seed germination of the foliar-resistant biotypes of *Erigeron canadensis* L. and *E. sumatrensis* Retz. Pesticide Biochemistry and Physiology. 1992;**44**:21

[136] Barr AR, Mansooji AM, Holtum JAM, Powles SB. The inheritance of herbicide resistance in *Avena sterilis* ssp. ludoviciana, biotype SAS. Proceedings of 1st International Weed Control Congress. 1992;**2**:70

[137] Christopher JT, Preston C, Powles SB. Malathion antagonizes metabolism-based chlorsulfuron resistance in *Lolium rigidum*. Pesticide Biochemistry and Physiology. 1994;**49**:172-182

[138] Barrett M. Metabolism of herbicides by cytochrome P450 in corn. Drug Metabolism and Drug Interactions. 1995;**12**:299

[139] Bishop T, Powles SB, Cornic G. Mechanism of paraquat resistance in *Hordeum glaucum*. II. Paraquat uptake and translocation. Australian Journal of Plant Physiology. 1987;**14**:539

[140] Konishi T, Shinohara K, Yamada K, Sasaki Y. Acetyl-CoA carboxylase in higher plants: Most plants other than gramineae have both the prokaryotic and eukaryotic forms of this enzyme. Plant Cell Physiology. 1996;**37**:117

[141] Tardif FJ, Preston C, Powles SB. Mechanisms of herbicide multiple resistance in *Lolium rigidum*. In: De Prado R, Jorrín J, García-Torres L, editors. Weed and Crop Resistance to Herbicides. Dordrecht, the Netherlands: Kluwer Academic Publishers; 1997. p. 117

[142] De Prado R, De Prado JL, Mendenez J. Resistance to substituted urea herbicides in Lolium rigidum biotypes. Pesticide Biochemistry and Physiology. 1997;**57**:126

[143] Gruys KJ, Biest-Taylor NA, Feng PCC, Baerson SR, Rodriguez DJ, You J, et al. Resistance of glyphosate in annual ryegrass (*Lolium rigidum*). II. Biochemical and molecular analyses. Weed Science Society of America Abstracts. 1999;**39**:82

[144] Lorraine-Colwill DF, Hawkes TR, Williams PH, Warner SAJ, Sutton PB, Powles SB, Preston C. Resistance to glyphosate in *Lolium rigidum*. Journal of Pesticide Science. 1999;**55**:486

[145] McAlister FM, Holtum JAM, Powles SB. Dinitroaniline herbicide resistance in rigid ryegrass (*Lolium rigidum*). Weed Science. 1995;**43**:55

[146] Woodworth AR, Bernasconi P, Subramanian MV, Rosen BA. A second naturally occurring point mutation confers broad based tolerance to acetolactate synthase inhibitors. Plant Physiology. 1996;**111**:105

[147] Turcsányi E, Darkó É, Borbély G, Lehoczki E. The activity of oxyradicaldetoxifying enzymes is not correlated with paraquat resistance in *Conyza Canadensis* (L.) Cronq. Pesticide Biochemistry and Physiology. 1998;**60**:1

[148] Guttieri MJ, Eberlein CV, Thill DC. Diverse mutations in the acetolactate synthase gene confer chlorsulfuron resistance in kochia (*Kochia scoparia*) biotypes. Weed Science. 1995;**43**:175

[149] Purba E, Preston C, Powles SB. The mechanism of resistance to paraquat is strongly temperature dependent in resistant *Hordeum leporinum* Link and *H. glaucum* Steud. Planta. 1995;196

[150] Sinning I, Koepke J, Michel H. Recent advances in the structure analysis of *Rhodopseudomonas viridis* mutants resistant to the herbicide terbutryn. In: Michel-Beyerle ME, editor. Reaction Centers of Photosynthetic Bacteria. Berlin: Springer-Verlag; 1990. p. 199

[151] Maneechote C, Preston C, Powles SB. A diclofop-methyl-resistant *Avena sterilis* biotype with a herbicide-resistant acetyl-coenzyme A carboxylase and enhanced metabolism of diclofop-methyl. Journal of Pesticide Science. 1995;**49**:105

[152] Hall LM, Stromme KM, Horsman GP, Devine MD. Resistance to acetolactate synthase inhibitors and quinclorac in a biotype of false cleavers (*Galium spurium*). Weed Science. 1998;**46**:390

[153] Holt JS, LeBaron HM. Significance and distribution of herbicide resistance. Weed Technology. 1990;**4**:141

[154] Saari LL, Maxwell CA. Target-site resistance for acetolactate synthase inhibitor herbicides. In: De Prado R, Jorrín J, García-Torres L, editors. Weed and Crop Resistance to Herbicides. Dordrecht: Kluwer Academic Publishers; 1997. p. 81

[155] Woodworth AR, Rosen BA, Bernasconi P. Broad range resistance to herbicides targeting acetolactate synthase (ALS) in a field isolate of *Amaranthus* spp. is conferred by a Trp to Leu mutation in the ALS gene (Accession No. U55852) (PGR96-051). Plant Physiology. 1996;**111**:1353

[156] Heap I. International survey of herbicide resistant weeds: Lessons and limitations. 1999 Brighton Crop Protection Conference: Weeds. Proceedings of an International Conference; 15-18 November 1999; Brighton, UK. 1999. p. 769

[157] Hirase K, Matsunaka S. Physiological role of the propanil hydrolyzing enzyme (aryl acylamidase I) in rice plants. Pesticide Biochemistry and Physiology. 1991;**41**:82

[158] Darmency H, Gasquez J. Inheritance of triazine resistance in *Poa annua*: Consequences for population dynamics. New Phytologist. 1981;**89**:487

[159] Frey JE, Müller-Schärer H, Frey B, Frei D. Complex relationship between triazine-susceptible phenotype and genotype in the weed *Senecio vulgaris* may be caused by chloroplast DNA polymorphism. Theoretical and Applied Genetics. 1999;**99**:578

[160] Sabet Zangeneh H, Mohammaddust Chamanabad HR, Zand E, Asghari A, Alamisaeid K, Travlos IS, Alebrahim MT. Study of fitness cost in three rigid ryegrass populations susceptible and resistant to acetyl-CoA carboxylase inhibiting herbicides. Frontiers in Ecology and Evolution. 2016;**4**:142

[161] National Research Council Committee on the Impact of Biotechnology on Farm-Level Economics and Sustainability. The Impact of Genetically Engineered Crops on Farm Sustainability in the United States. Washington, DC: The National Academies Press; 2010

[162] Holt JS. History of identification of herbicide resistant weeds. Weed Technology. 1992; **6**:615-620

[163] Duke SO, Powles SB. Mini-review. Glyphosate: A once-in-a-century herbicide. Pest Management Science. 2008;**64**:319-325

[164] Baker HG. Characteristics and modes of origin of weeds. In: Baker HG, Stebbins GL, editors. The Genetics of Colonizing Species. New York: Academic Press; 1965. pp. 147-172

[165] Heap I, LeBaron H. Introduction and overview of resistance. In: Powles SB, Shaner DL, editors. Herbicide Resistance and World Hrains. Boca Raton, FL: CRC Press; 2001. pp. 1-22

[166] Holm LG, Doll J, Holm E, Pancho JV, Herberger JP. World Weeds: Natural Histories and Distribution. New York, NY: John Wiley and Sons, Inc.; 1997

[167] Holm LG, Plucknett DL, Pancho JV, Herberger JP. The World's Worst Weeds: Distribution and Biology: East-west Center. Honolulu, HI: University Press of Hawaii; 1977

[168] Harper JL. The evolution of weeds in relation to resistance to herbicides. In: Proceedings of the 3rd British Weed Control Conference. 1956. pp. 179-188

[169] Gressel J, Segel LA. The paucity of plants evolving genetic resistance to herbicides: Possible reasons and implications. Journal of Theoretical Biology. 1978;**75**:349-371

[170] Maxwell BD, Roush ML, Radosevich SR. Predicting the evolution and dynamics of herbicide resistance in weed populations. Weed Technology. 1990;**4**:2-13

[171] Holt JS, Welles SR, Silvera K, Heap IM, Heredia SM, Martinez-Berdeja A, et al. Taxonomic and life history bias in herbicide resistant weeds: Implications for deployment of resistant crops. PLOS ONE. 2013;**8**(9):e71916

[172] Cerdeira AL, Duke SO. The current status and environmental impacts of glyphosate-resistant crops: A review. Journal of Environmental Management. 2006;**35**:1633-1658

[173] Duke SO. Herbicide-resistant crops - background and perspectives. In: Duke SO, editor. Herbicide- resistant Crops. Boca Raton: Lewis; 1996. pp. 1-12

[174] Mazur BJ, Falco SC. The development of herbicide resistant crops. Annual Review of Plant Physiology and Plant Molecular Biology. 1989;**40**:441-470

[175] Stalker DM, Kiser JA, Baldwin G, Coulombe B, Houck CM. Cotton weed control using the BXN system. In: Duke SO, editors. Herbicide-resistant Crops. Boca Raton: Lewis; 1996. pp. 93-105

[176] Padgette SR, Re DB, Barry GF, Eichholtz DE, Delannay X, Fuchs RL, Kishore GM, Fraley RT. New weed control opportunities: Development of soybean with Roundup Ready gene. In: Duke SO, editor. Herbicide-resistant Crops. Boca Raton: Lewis; 1996. pp. 53-84

[177] Wells BH. Development of glyphosate tolerant crops into the market. Brighton Crop Protection Conference: Weeds. 1995. 787-790

[178] Inui H, Ueyama Y, Shiota N, Ohkawa Y, Ohkawa H. Herbicide metabolism and cross tolerance in transgenic potato plants expressing human CYP1Al. Pesticide Biochemistry and Physiology. 1999;**l64**:33-46

[179] Degryse E, Freyssinet G, Lebrun M, Sailland A. Mutated 5-enol-pyruvylshikimate-3-phosphate synthase gene coding for this protein and transformed plants containing this gene. 1997. Patent no. WO 97/04103

[180] HollaKnder-Czytko H, Johänning D, Meyer HE, Amrhein N. Molecular basis for the overproduction of 5-enolpyruvylshikimate-3-phosphate synthase in a glyphosate-tolerant cell suspension culture of *Corydalis sempervirens*. Plant Molecular Biology. 1988;**11**:215-220

[181] Rogers SG, Brand LA, Holder SB, Sharps ES, Brackin MJ. Amplification of the *aroA* gene from Escherichia coli results in tolerance to the herbicide glyphosate. Applied and Environmental Microbiology. 1983;**46**:37-43

[182] Wang Y, Jones JD, Weller SC, Goldsbrough PB. Expression and stability of amplified genes encoding 5-enolpyruvylshikimate-3-phosphate synthase in glyphosate-tolerant tobacco cells. Plant Molecular Biology. 1991;**17**:1127-1138

[183] Lee KY, Townsend J, Tepperman J, Black M, Chui CF, Mazur B, Dunsmuir P, Bedbrook J. The molecular basis of sulfonylurea herbicide resistance in tobacco. EMBO Journal. 1988;**7**:1241-1248

[184] Hauser CR, Oeda K, et al. Isolation and characterization of a mutant protoporphyrinogen oxidase gene from Chlamydomonas reinhardtii conferring resistance to porphyric herbicides. Plant Molecular Biology. 1998;**38**:839-859

[185] Saari LL, Mauvais CJ. Sulfonylurea herbicide resistant crops. In: Duke SO, editor. Herbicide Resistant Crops. Boca Raton: Lewis; 1996. pp. 127-142

[186] McBride KE, Svab Z, Schaaf DJ, Hogan PS, Stalker DM, Maliga P. Amplification of a chimeric Bacillus gene in chloroplasts leads to an extraordinary level of an insecticidal protein in tobacco. Bio/Technology. 1995;**13**:362-365

[187] Oey M, Lohse M, Kreikemeyer B, Bock R. Exhaustion of the chloroplast protein synthesis capacity by massive expression of a highly stable protein antibiotic. Plant Journal. 2009;**57**:436-445

[188] Ruf S, Karcher D, Bock R. Determining the transgene containment level provided by chloroplast transformation. Proceedings of the National Academy of Sciences of the United States of America. 2007;**104**:6998-7002

[189] Thyssen G, Svab Z, Maliga P. Exceptional inheritance of plastids via pollen in *Nicotiana sylvestris* with no detectable paternal mitochondrial DNA in the progeny. Plant Journal. 2012;**72**:84-88

[190] Maliga P. Plastid transformation in higher plants. Annual Review of Plant Biology. 2004;**55**:289-313

Chapter 6

# Predictions for Weed Resistance to Herbicides in Brazil: A Botanical Approach

André Andres, Germani Concenço, Fábio Schreiber,
Dirceu Agostinetto, Leandro Vargas, João Behenck,
Giovanni Antoniaci Caputo and
Ygor Sulzbach Alves

Additional information is available at the end of the chapter

http://dx.doi.org/10.5772/intechopen.68336

**Abstract**

The intensive use of herbicides in agriculture has led to the appearance of resistant weed biotypes. Resistance is the inherited ability of a plant to survive following application of an herbicide dose which should be lethal. Morphophysiological weed traits help defining the risk to evolve resistance. These traits are not exclusive to the species but may be innate to botanical order, family, or genus. Four reference countries were screened about the nature of resistance—Australia, Canada, France, and the United States—and the data were used for predictions in the Brazilian scenario. Most weed species with resistant biotypes in the reference countries seem to be native to the continent. The most important botanical families with resistant biotypes in the reference countries were also among the first ones to develop resistance in these countries. There was a predominance of C3 species over C4 in the number of plant species with resistant biotypes in the reference countries. In Brazil, three orders are considered as high risk (Gentianales, Lamiales, and Solanales), besides the six already present. Furthermore, eight botanical families present superior risk to evolve resistance and for five of them (Caryophyllaceae, Polygonaceae, Rubiaceae, Convolvulaceae, and Solanaceae), resistance cases have not been reported to date in Brazil.

**Keywords:** weed species, botanical traits, herbicide, plant selection, carbon metabolism

## 1. Introduction

The successive and intensive cultivation of the same crop species in Brazil, with practically no crop rotation, is leading to an increase in the presence of weeds [1]. One should emphasize that the term "weed" had effectively no botanical meaning, since by the classical definitions, a plant can be a "weed" in a given situation while it may be desirable in another. Distinct plant species included into the same botanical order may be considered as weed or desirable. Furthermore, the same species may be considered as a "weed" into an arable field while it can be desirable in gardens, for instance, where they are usually considered as "beneficial weeds" [2].

The use of herbicides for weed control in Brazilian agriculture has increased significantly in the last years, due to a series of factors, such as the growing difficulty to find human labor for manual weeding and the excessive damage caused to plants when adopting in-crop mechanical control [2]. Moreover, the chemical control represents an easy and efficient approach for weed control and therefore farmers are most prone to use this method despite of the other weed suppression strategies [1].

Weed species that have been indirectly selected for adverse conditions obtain their vital elements more efficiently by extracting water, nitrogen, phosphorus, and potassium, respectively, four, five, three, and six times more than crop plants [3]. Thereby, due to their ability to compete for environmental resources with cultivated plants, it is essential to eliminate them from cropping fields. Considering also that usually for every crop there is specific companion weeds [4], weed control based solely on herbicides tends to reduce quickly their efficiency due to plant selection which become resistant or tolerant to these compounds [1]. Compared to other pests, weeds have longer reproduction cycles [5] and produce propagules which survive in soil for several years [1]. These factors contribute to the relatively slow evolution of resistant weeds compared to other pests.

Weed resistance to herbicides is defined as the inherited ability of a plant to survive following application of the commercially used dose of the herbicide recommended for its control. This dose, in regular conditions, should be able to control that weed species [2]. There are several factors responsible for selecting resistant weed biotypes, as the selection pressure imposed by the herbicide [1]. Herbicides differ in the risk that they present to select a given resistant weed biotype, and this depends, among other aspects, on its specificity in terms of point of action into the plant; more specifically the local of action [6].

For instance, the herbicide 2,4-D, is a synthetic auxin used continuously since 1948 and the first case of resistance to this compound was reported only in 1957 [6]. Currently, 40 years later, resistance cases to synthetic auxins was documented only for 14 species in 11 countries [1]. Thus, herbicides from this mode of action and other inhibitors, such as Protox and EPSPs inhibitors, are examples of "low-risk" herbicides for resistance evolution [1, 7]. On the other hand, herbicides like the ones included in the acetolactate synthase (ALS)-inhibiting group are considered as "high risk" for herbicide resistance evolution. This classification for a new resistance case to appear is based on the location in which the herbicide acts into the plant and other aspects [7]; more specifically, its mechanism of action, as a single mutation into the plant, could turn it resistant to the herbicide [6]. The time required for the appearance of the first resistant biotype to commonly used herbicides worldwide from their introduction in the market are shown in **Table 1**.

The history and concepts about weed resistance have been widely explored in the literature including topics dealing with the mechanisms conferring resistance and herbicide traits which most easily select resistant biotypes. Although weed resistance is a well-known problem and is

http://dx.doi.org/10.5772/intechopen.68336

| Herbicide or mode of action | Introduction to the market | First resistance report | Introduction to first case (years) | Location |
|---|---|---|---|---|
| 2,4-D | 1948 | 1957 | 9 | USA and Canada |
| Triazines | 1959 | 1970 | 11 | USA |
| Propanil | 1962 | 1991 | 29 | USA |
| Paraquat | 1966 | 1980 | 14 | Japan |
| EPSPs inhibitors | 1974 | 1996 | 22 | Australia |
| ACCase inhibitors | 1977 | 1982 | 5 | Australia |
| ALS inhibitors | 1982 | 1984 | 2 | Australia |

Source: adapted from Agostinetto and Vargas [6].

**Table 1.** Time required for appearance of the first resistant weed biotype following introduction of a new herbicide mechanism of action into the market.

relatively characterized, its occurrence is constantly increasing in a worldwide basis. To assist researchers to keep updated about herbicide resistance spread around the world, there is a website, www.weedscience.org, which is used as a platform for researchers to register the new cases of weed resistance [8]. This website is maintained by the Global Herbicide Resistance Action Committee and CropLife International, and it is an open access tool. The basic worldwide data about weed resistance used in the present study were obtained from that site, which were used with permission from the owners. Further data to botanically characterize the weed species listed on the WeedScience website were obtained from specialized literature.

Besides herbicide risk and frequency of application to the field, which are already well studied in the literature related to the weed science, other plant traits could turn them resistant to herbicides. Morphophysiological characteristics as dormancy behavior, number of seeds produced, annual distribution of emergence, and several others [1] can maximize the chance for the occurrence of weed species in the fields at the time of herbicide application, thus exposing them to the selection pressure imposed by the herbicide [7]. These traits can be studied as not being exclusive for the plant species but a characteristic innate to the botanical order, family, or genus of the weed species with resistant biotypes. Supposing this relationship exists, plants which are most closely related to resistant species could also be most prone to evolve resistance. The present study is based upon this hypothesis.

In order to have a wider comprehension about the path resistance takes into the botanical classification of weed species, from the appearance of the first resistant species to the current situation of resistance in Brazil and its most probable future, four reference countries were selected to serve as background for understanding the Brazilian context of weed resistance and new future resistance cases. The countries with higher number of resistant weed biotypes were first selected; in the second stage of selection were selected among these countries two that represented the American Continent, where Brazil is located; one which represented Europe, from where some weed species are known to be introduced in Brazil, and one to represent Oceania, where the climate is more alike to the observed in several regions of Brazil. In this context, the following countries were selected to be studied: Australia, Canada, France, and the United States.

## 2. Chemical classification of resistance

Taking into account the herbicides' mechanisms of action, it is possible to clearly observe a predominance in cases of resistance to the application of the acetolactate synthase (ALS)-inhibiting herbicides in Australia, Canada, and the United States (**Figure 1**), followed by resistance to PSII-inhibiting herbicides in France. The number of resistance cases to PSII-inhibiting herbicides is also high in the other countries, rating this mechanism of action in the second place in number of reported resistance cases in the United States and Canada. In Australia, the number of resistance cases to ACCase and EPSPs herbicides was in the second place, followed by the PSI and PSII herbicides (**Figure 1**).

In comparison to other mechanisms of action, resistance cases to ACCase-inhibiting herbicides were very important in all countries. Moreover, EPSP herbicides were also important in Australia and the United States, but, in general, the number of resistance cases to this mechanism of action was smaller in Canada and France (**Figure 1**). Overall, there were

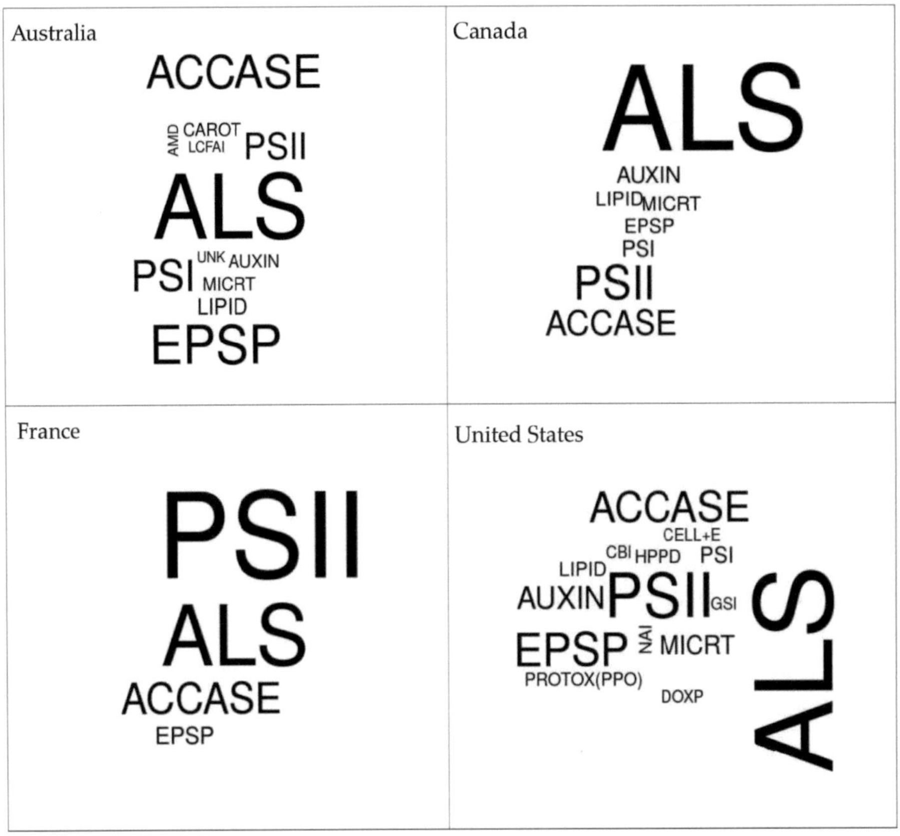


**Figure 1.** Wordcloud for the occurrence of weeds resistance as a function of mechanism of action in the reference countries. The scale of the font represents the importance of the mechanism compared to the others in the same figure. Source: adapted from Heap [8].

weeds resistant to 12, 8, 4, and 14 herbicidal mechanisms of action, respectively, in Australia, Canada, France, and the United States (**Figure 1**).

To date (November 2016) in Brazil, there are cases of weeds resistant to five mechanisms of action (**Figure 2**). Furthermore, the majority of these cases is associated with ALS herbicides. In the second place comes the resistance against EPSPs, followed by ACCase, Auxin, and Protox herbicides. Compared to other countries, one may observe that the order of importance of herbicide mechanisms of action in Brazil resembles more closely to the Australia and the United States context (**Figure 1**). The number of resistance cases to ACCase herbicides is similar to Canada; however, resistance to EPSP herbicides is not as important in that country as compared to Brazil. Moreover, the number of resistance cases to each mechanism of action in France was the one that contrasted the most from Brazil (**Figure 1**).

France is characterized by growing large areas of barley, to supply the demand of breweries [9, 10], and oat. Maize is also a common crop in France, where the French production of these crops, potatoes and sugarbeets, helps to meet the demand of these products in Europe [9]. In the cooler regions of France, apples are cultivated, as well as grapes for the production of wine [10]. Other French crops mostly include plums, tomatoes, and peaches [9].

The Brazilian agriculture differs from French crops as it is based mostly on maize, wheat, rice, soybeans, orange (*in natura* and juice), sugarcane (including sugar and ethanol), cotton, cassava, coffee, potatoes. Fruits such as grapes, apples, bananas, mangoes, melons, tobacco, papaya, and

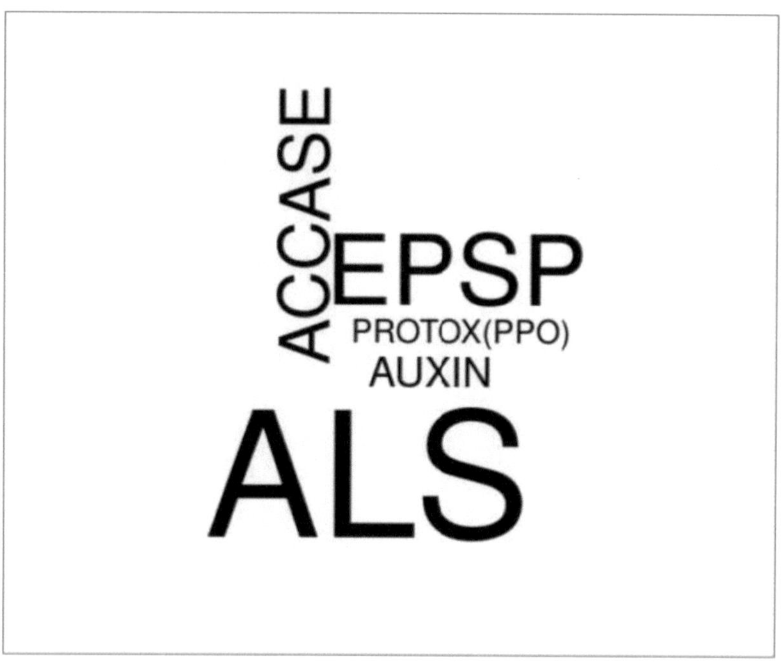


**Figure 2.** Wordcloud for the occurrence of weeds resistance as a function of mechanism of action in Brazil. The scale of the font represents the importance of the mechanism compared to the others in the same figure. Source: adapted from Heap [8].

pulp are also cultivated. Moreover, the paper industry is also important in the country [11]. As the world's sixth largest economy, Brazil ranks third among the world's major agricultural exporters and fourth for food products, being the world's largest producer and exporter of products such as soybean, coffee, sugarcane, orange juice, meat, and tobacco [12].

Thus, there is a great difference between the major crops grown in France and Brazil, which is probably the cause for a distinct herbicide demand and, as consequence, the difference in nature of resistance cases between these two countries.

## 3. Botanical classification of resistance

The botanical classification of life forms is very often, if not always, a challenge for agronomists. In order to understand how nature is reacting to the heavy load of herbicides continuously thrown into the environment, first there is a need to briefly understand the botanical classification and how plants are grouped.

The Biological Classification—or Taxonomic Rank—describes the level of a group of organisms into the taxonomic hierarchy [13]. The main taxonomic ranks are domain, kingdom, division, class, order, family, genus, and species, all of them with an internal classification prefixed by "sub" (subclass, subfamily, etc.) [14]. Other classification levels into each section may exist, but this is beyond an agronomist's point of view and will not be discussed in the present study.

### 3.1. Botanical order

In botany, "order" is a taxonomic rank located between "class" and "family," grouping plants with similar traits at a certain degree [14]. Even though several botanical orders exist, most weed species should be classified into approximately 20 orders. **Figure 3** depicts weed species distribution into orders, from combined data of the five studied countries (Australia, Brazil, Canada, France, and the United States).

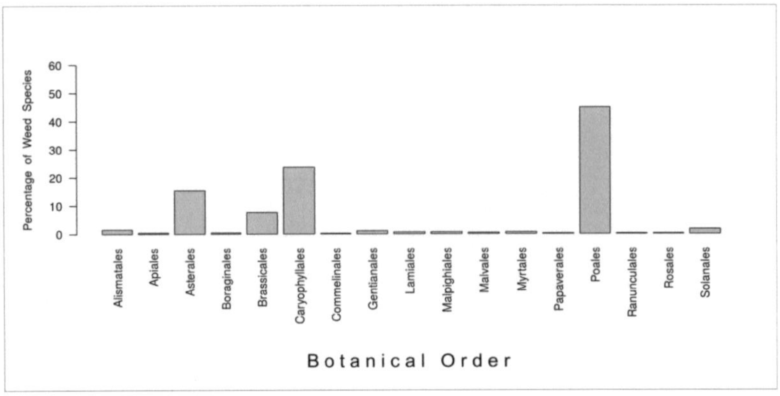


**Figure 3.** Percentage of resistant weed species by botanical order, with pooled data from the five studied countries. Source: adapted from Heap [8].

The great majority of weed species resistant to herbicides are included into the orders Poales, Caryophyllales, Asterales, and Brassicales (**Figure 3**). For the four reference countries (**Figure 4**), the order Poales was the predominant one, as also seen for the overall order data (**Figure 3**). In Canada, the importance of this order was shared with Caryophyllales, which was the second most important resistant weed group in France. Asterales was of importance also in Australia and the United States while it was of secondary importance in Canada and France (**Figure 4**). Overall, 10, 8, 6, and 12 orders including resistant weed species were identified in Australia, Canada, France, and the United States, respectively.

In Brazil, the most important weed species are included into the botanical orders Poales, Asterales, and Caryophyllales (**Figure 5**). In general, these findings are according to the data observed for the four reference countries (**Figure 4**), where these three botanical orders also tended to predominate.

There are five, four, three, and six botanical orders, respectively, in Australia, Canada, France, and the United States, with resistant biotypes (**Figure 4**), which are still absent in Brazil (**Figure 5**). Among the plant orders with resistant biotypes, Solanales is present in the four reference countries, whereas Lamiales and Gentianales are present in Australia and Canada (**Figure 4**). The order Solanales includes botanical families with important weed species in Brazil like Solanaceae and Convolvulaceae; Lamiales includes the families Plantaginaceae and Lamiaceae, while Gentianales includes the family Rubiaceae [14].

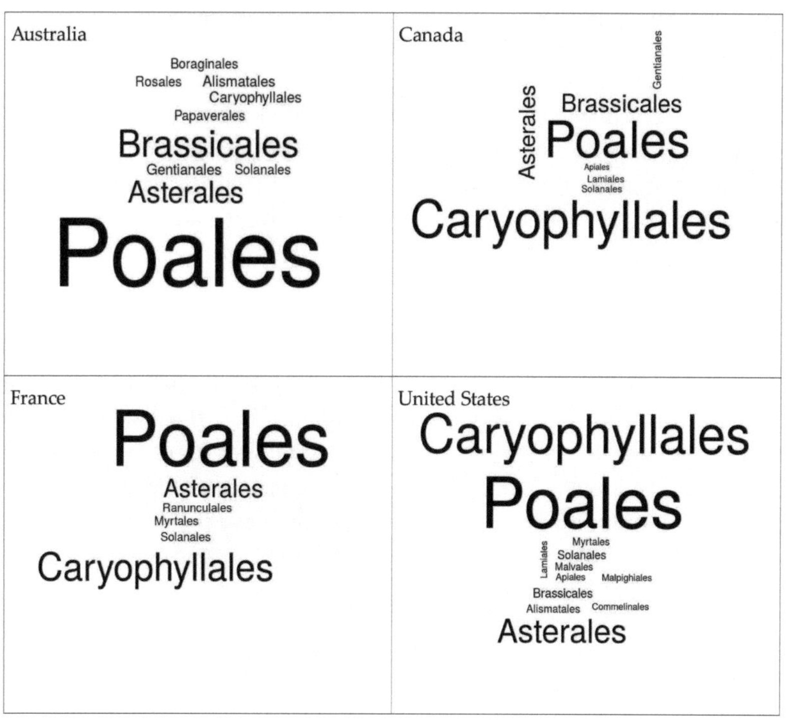


**Figure 4.** Wordcloud for the occurrence of weed resistance as a function of botanical order in the reference countries. The scale of the font represents the importance of the order compared to the others in the same figure. Source: adapted from Heap [8].

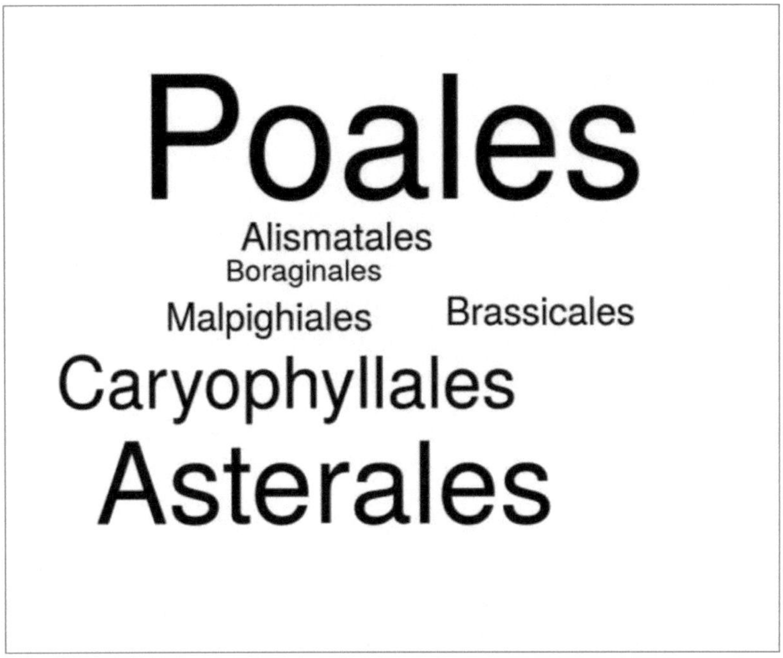


**Figure 5.** Wordcloud for the occurrence of weed resistance as a function of botanical order in Brazil. The scale of the font represents the importance of the order compared to the others in the same figure. Source: adapted from Heap [8].

It should be emphasized that in the present study, the classification of weeds in botanical orders is restricted to those species considered as "weed" in the agricultural context; this does not mean at all major number of plant species included in that order are most prone to become a weed. This relationship is yet to be established, supposing it exists.

### 3.2. Botanical family

Resistant weeds grouped by botanical family (**Figure 6** and **Figure 7**) showed Poaceae as the major family with resistant species for all studied countries, including Brazil. In Australia, Brassicaceae and Asteraceae families were in the second and third places, respectively; in Canada, Amaranthaceae was the family with the most number of plant species with resistant biotypes to herbicides followed by Brassicaceae, Asteraceae, and Chenopodiaceae (**Figure 6**). In France, Asteraceae, Amaranthaceae, and Polygonaceae were also important botanical families in number of weeds with resistant biotypes, and in the United States, Amaranthaceae and Asteraceae were highlighted after Poaceae (**Figure 6**).

In Brazil, Asteraceae and Amaranthaceae were the predominant families of plants with resistant weed biotypes after Poaceae (**Figure 7**).

Compared to the reference countries, from one point of view, there is danger in Brazil—under the current panorama of herbicides use and weed management—for an increase in the

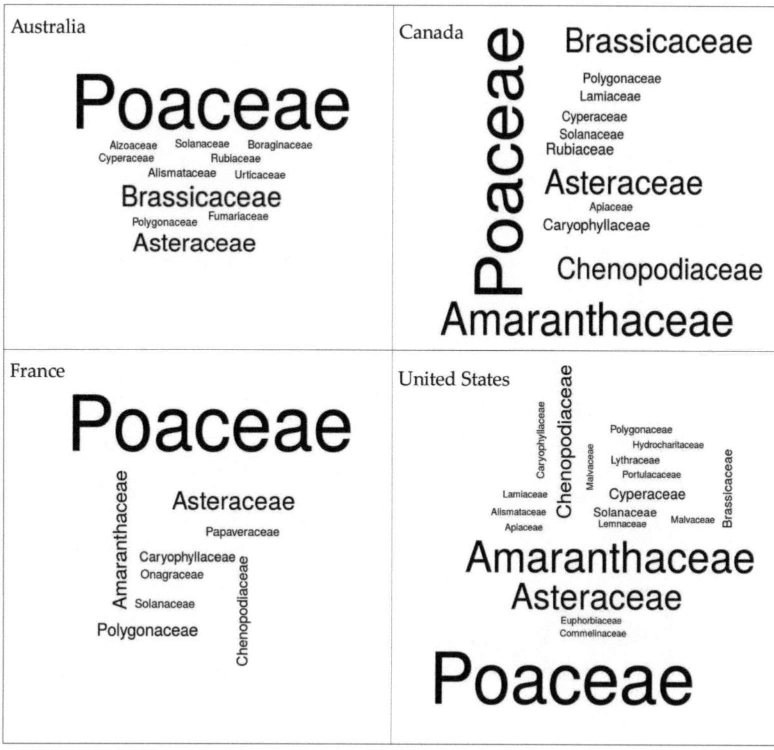


**Figure 6.** Wordcloud for the occurrence of weed resistance as a function of botanical family in the reference countries. The scale of the font represents the importance of the family compared to the others in the same figure. Source: adapted from Heap [8].

number of resistant plant biotypes mainly from the families Asteraceae, Amaranthaceae, and Brassicaceae, which are significant in the reference countries and, at the same time resistant, biotypes were already reported in Brazil.

Second, there is also a great chance for the appearance of resistant weed biotypes from families which are absent in the current Brazilian scenario, but which have great importance in all four reference countries, like Polygonaceae and Solanaceae, or in three out of the four reference countries like Chenopodiaceae and Caryophyllaceae (**Figure 6**).

A third scenario leads to the increasing number of multiple resistance or the appearance of resistance to a second herbicidal mechanism of action in weed species which are already resistant to a given mechanism of action [6]. This is concerning due to the nature of the new technologies of crop tolerance to herbicides, the so-called "all-in-one" tolerance; crops will have tolerance to more than one herbicide mechanism of action [15]. Thus, weeds will have to "tolerate" or "resist" to most of the herbicides associated to each technology, in order to prevail in arable fields.

One should note that there was an initial attempt to predict future cases of weed resistance in Brazil [16], by analyzing the herbicidal mechanism of action to which some plant

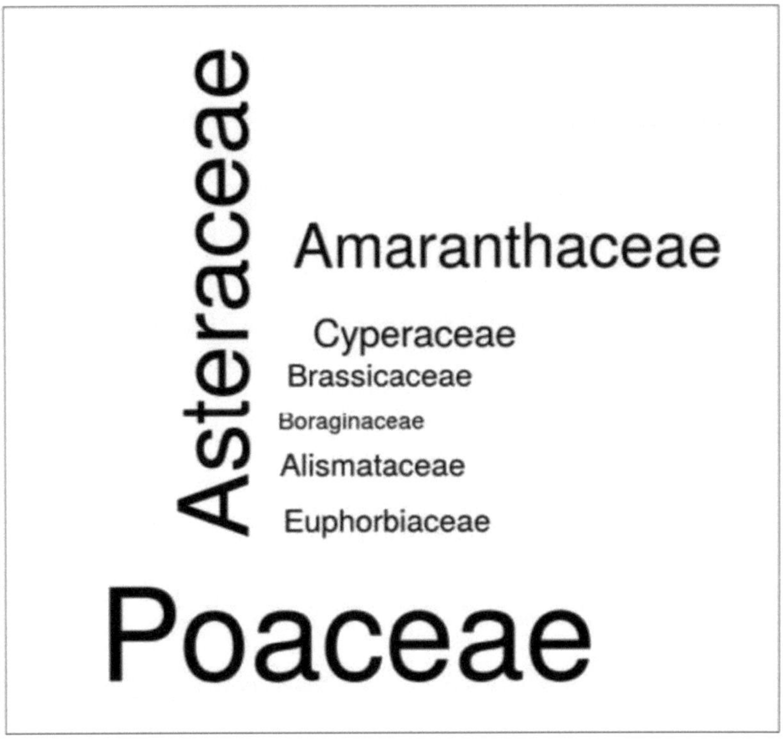


**Figure 7.** Wordcloud for the occurrence of weed resistance as a function of botanical family in Brazil. The scale of the font represents the importance of the family compared to the others in the same figure. Source: adapted from Heap [8].

species evolved resistance in some parts of the world, and relating the risk for new cases to the adoption of such herbicides in Brazilian agriculture. This analysis [16] was, however, excluded in the updated version of the same book [15] but may be considered as complementary to the present study, even being outdated by some degree.

When the evolution of appearance of families with resistant biotypes was analyzed by countries (**Figure 8**), it was observed that the most important botanical family in number of resistance cases is Poaceae (**Figure 6** and **Figure 7**), which was the first to appear in Australia, in 1982; the second in France and the United States in 1978 and 1970, respectively; and the sixth botanical family to have resistant biotypes in Canada (**Figure 8**).

The other botanical families with resistant biotypes with great importance that were reported in the reference countries (**Figure 6**) are listed in the inset table in **Figure 8**. It is important to note that the most important botanical families were also, in general terms, between the first ones to appear in the respective countries. This leads to the hypothesis that these families are of relatively recent evolutionary origin [5]. There is evidence that the preponderance of weeds from relatively recent evolutionary origin indicates the trend to an increasing of troublesome, highly adaptable weeds in agriculture [5, 7]. An example is the botanical family Asteraceae [7], which already has resistant biotypes in the five countries studied here (**Figure 5** and **Figure 7**). Thereby, in the years to come, Brazil may experience an increase in the occurrence of weed species from families with recent evolutionary origin.

http://dx.doi.org/10.5772/intechopen.68336

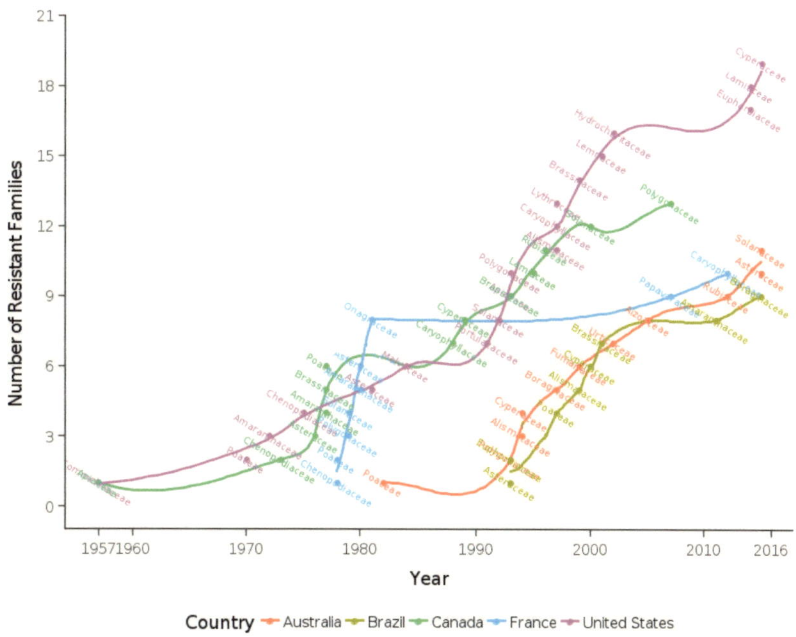


**Figure 8.** Evolution of resistant botanical families by year and country, with an inset table showing the number of families in each reference country with resistant biotypes, and the number of these families that are present in Brazil, with or without resistant biotypes. Source: adapted from Heap [8].

**Figure 9** illustrates the geographic distance between Brazil and the reference countries used in this study. In general terms, half or less than half of the families present in the reference countries are also present in Brazil with resistant biotypes. This may lead to the assumption that there is still plenty of species to evolve resistance in Brazil, supposing that farmers and technicians will keep relying heavily on the chemical weed control, in absence of alternative weed management techniques.

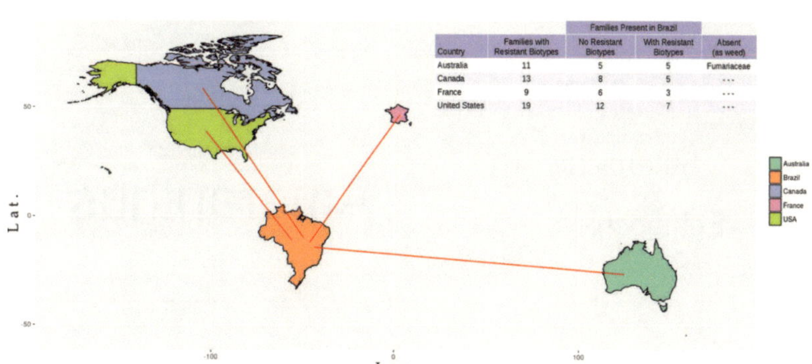


| Country | Families with Resistant Biotypes | Families Present in Brazil: No Resistant Biotypes | Families Present in Brazil: With Resistant Biotypes | Absent (as weed) |
|---|---|---|---|---|
| Australia | 11 | 5 | 5 | Fumariaceae |
| Canada | 13 | 8 | 5 | ... |
| France | 9 | 6 | 3 | ... |
| United States | 19 | 12 | 7 | ... |

**Figure 9.** Correlation in the occurrence of botanical families of weeds between Brazil and the reference countries, and its proportion of families which already present resistant biotypes in Brazil. Source: adapted from Heap [8].

### 3.3. Botanical genera

The botanical genera with resistant species in the reference countries are shown in **Figure 10**. In Australia, *Lolium* was the predominant genus in number of resistant weed species; in Canada and the United States, *Amaranthus* is the most important one; in France, there is no predominant genus with the most cases of resistant weed biotypes being *Avena, Amaranthus, Lolium, Setaria,* and *Echinochloa* similar in importance (**Figure 10**). In Australia, *Raphanus, Bromus, Hordeum, Avena,* and *Sisymbrium* are the second most important group of genera with resistant weed biotypes; in Canada, *Setaria* and *Avena* are also in the second group. In the United States, a great number of botanical genera with resistant weed biotypes occur, but *Echinochloa, Conyza, Poa, Setaria, Kochia, Ambrosia,* and *Lolium* may be highlighted in a second group of importance, following *Amaranthus*.

In Brazil, 19 genera with resistant biotypes are reported (**Figure 11**), where *Amaranthus, Conyza,* and *Bidens* are the most important ones, followed by *Digitaria, Lolium,* and *Echinochloa* in the second group. A third group includes *Sagittaria, Euphorbia, Eleusine, Cyperus,* and *Raphanus*.

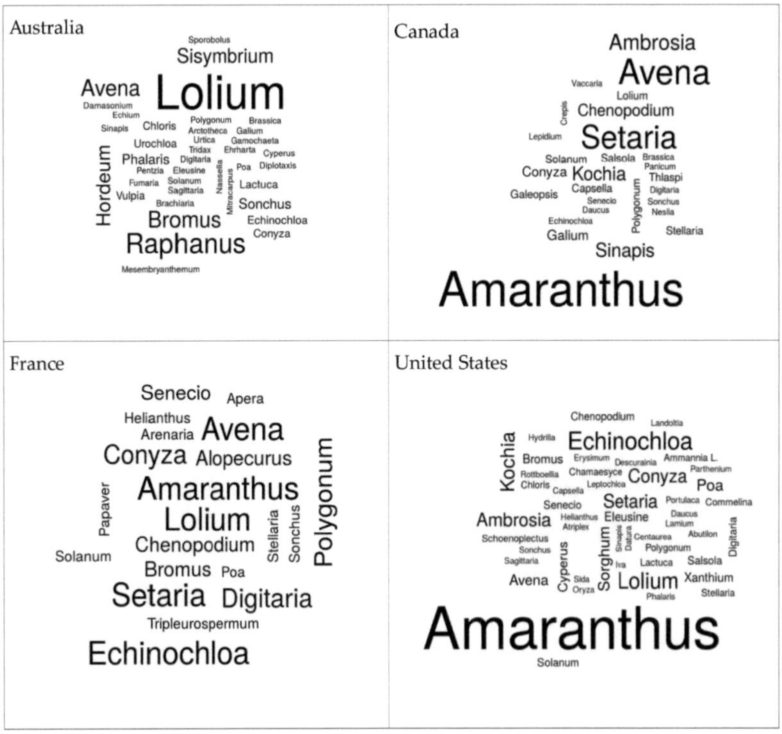


**Figure 10.** Wordcloud for the occurrence of weed resistance as a function of botanical genus in the reference countries. The scale of the font represents the importance of the genus compared to the others in the same figure. Source: adapted from Heap [8].

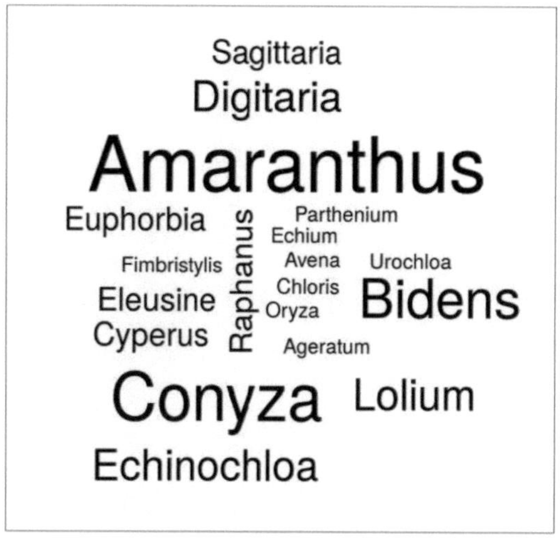


**Figure 11.** Wordcloud for the occurrence of weed resistance as a function of botanical family in Brazil. The scale of the font represents the importance of the genus compared to the others in the same figure. Source: adapted from Heap [8].

## 4. Botanical class and carbon metabolism pathway

As mentioned earlier, plant taxonomy is not static, and from time to time some adaptations are proposed to plant nomenclature by different authors [17], trying to adjust plant classification under the light of new evolutionary evidences or simply aiming to rearrange previous taxonomic trees. Angiosperms are, in a free definition, plants with flowers whose seeds are protected in fruits [13, 18]. Along the history, different plant classification systems were proposed, which can be roughly divided into three groups: (1) artificial systems, based on superficial features; (2) natural systems, based on form relationships; and (3) phylogenetic systems, based on evolutionary and genetic relationships [18].

The artificial systems are very old, based usually on a single character, and have been used as example by Theophrastus (370 285 BC) and Linnaeus (1707 1778 AD); natural systems were based on a set of botanical characters, being used in the eighteenth and nineteenth centuries; examples are the classification systems of Jussieu and Bentham & Hooker [19]. Among the phylogenetic systems, Cronquist [20], later reviewed in 1988 [21], is one of the most used, and it divides flowering plants into two classes: (1) Magnoliopsida (dicotyledons, dicots) and (2) Liliopsida (monocotyledons, monocots). With no intention to start a war among plant taxonomists and considering the division into these two classes is the most common in the weed science, we grouped resistant plant species into dicots and monocots (**Figure 12** and **Figure 13**).

Australia has 34 dicot species with resistant biotypes, while 54 monocot weed species had at least one resistant biotype (**Figure 12**). Monocot species were also the majority in France, where 20 dicots and 26 monocots had at least one resistant biotype. In the northern region of the American continent, dicots predominated among the species with resistant biotypes; 59 and 108 dicots contrasted with 31 and 87 monocots, in Canada and the United States, respectively (**Figure 12**).

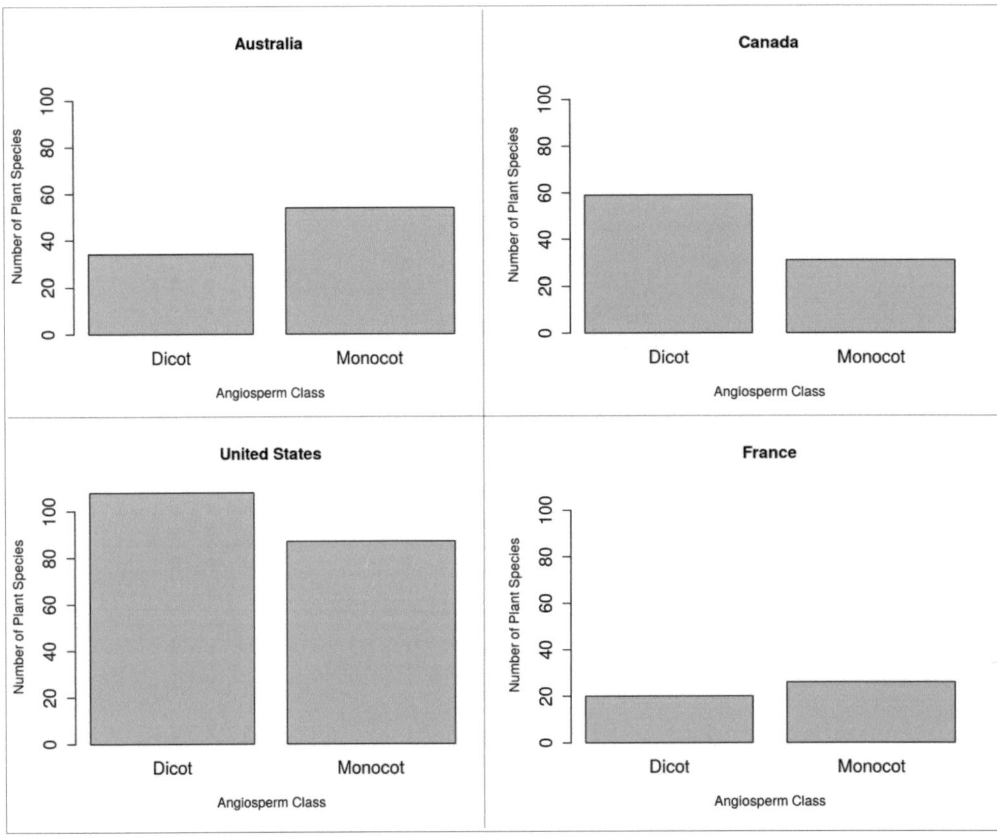


**Figure 12.** Occurrence of weed resistance as a function of botanical class in the reference countries. Source: adapted from Heap [8].

When weed species with resistant biotypes were grouped in the reference countries by the carbon metabolism (C3, C4, intermediary/hybrid/unknown) (**Figure 13**), there was a clear predominance of C3 species over C4 in the number of plant species with resistant biotypes for all countries; 75, 49, 115, and 25 weed species with at least one biotype resistant to herbicides were C3, while only 5, 31, 71, and 14 were C4, respectively, for Australia, Canada, the United States, and France (**Figure 13**).

Plant species with carbon metabolism by the C4 cycle, in evolutionary terms are derived from the C3 cycle [22]; furthermore, although it is generally claimed that C4 plant species are most widely distributed in warmer and dry environments compared to C3 plants, this is not remarkable since C4 plants evolved to optimize carbon fixation in low-C environments, and not essentially to resist to water stresses as usually believed [23]. In fact, C4 plants may be equally or even more sensitive to water stress than C3 species, in spite of the greater water use efficiency of C4 plants [24].

In Brazil, where the majority of the arable territory is located in warm climates with mild winters, there were no significant differences in the proportion of dicots (22) and monocots (20) with resistant biotypes (**Figure 14**). In North America (Canada and the United States), dicot

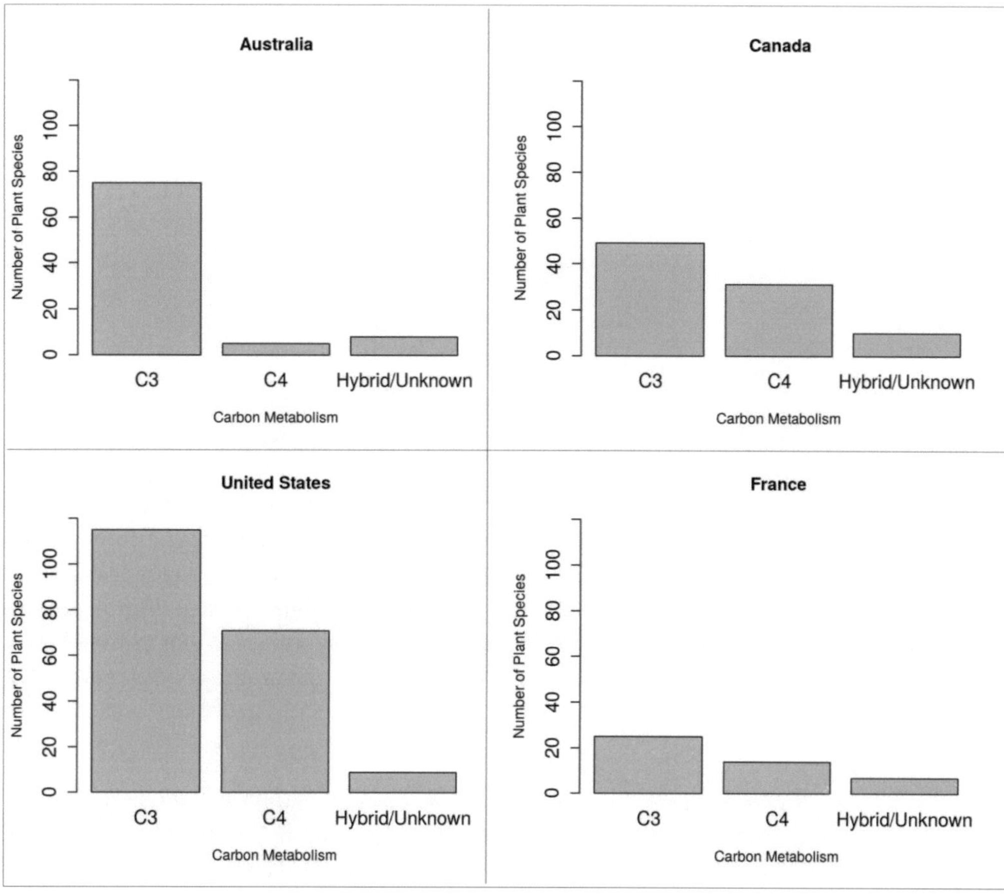


**Figure 13.** Occurrence of weed resistance as a function of carbon metabolism pathway in the reference countries. Source: adapted from Heap [8].

weed species with resistant biotypes predominated, while in France and Australia, monocots tended to predominate (**Figure 13**).

Dicot species may have advantages over monocots. With no intention to differentiate these two groups of plants, some traits from each group may be highlighted: first, the vascular bundle of dicots may allow flow of higher volumes of sap to and from leaves, as well as up and down into the plant compared to monocots; second, the stronger vascular bundle could allow dicots to resist stronger water potentials, which could be advantageous in both rich and scarce water environments; third, the two cotyledons could allow for higher photosynthesis rates to dicots, which would depend less on the seed stored energy to form its initial leaf area, increasing their chance of survival [22]. These facts could help explain why dicots were superior to monocots in Canada and the United States. On the other side, bulliform cells which are present in many monocots—not only in grasses—may help avoiding stress by excessive light incidence in low latitude environments [23, 24], which could turn it into a big advantage for some groups of monocots in tropical agriculture.

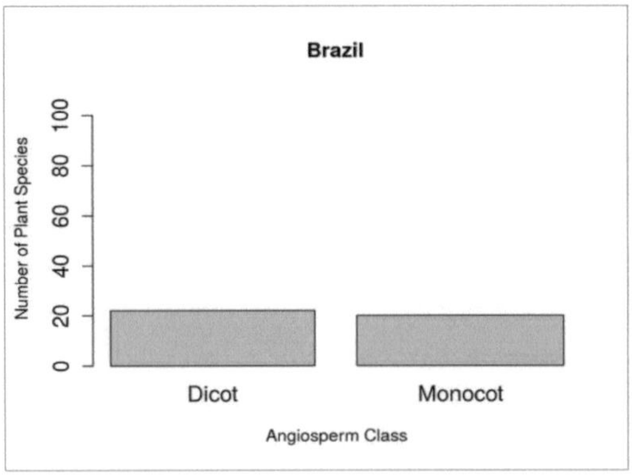


**Figure 14.** Occurrence of weed resistance as a function of botanical class in Brazil. Source: adapted from Heap [8].

There was also no difference in the proportion of plants as a function of the carbon metabolism pathway (**Figure 15**). Ehleringer and Monson [23] report that in anthropogenically altered environments, C4 plants are usually not so advantageous over C3. In the reference countries (**Figure 13**), most weed species with resistant biotypes were C3, but in Brazil (**Figure 15**) this difference was not remarkable. By considering this, one may hypothesize that in Brazil most of the arable lands are as intensely explored as consequence of the anthropogenic effect, that it led C4 plants to almost totally lose their superior potential compared to C3 weeds in the same environment.

When the data of angiosperm class is crossed with the data of carbon metabolism pathway (**Figure 16**), visually there appears to be little to no relationship between these factors; but when we apply a $X^2$ test to the data *Ang.Class* x *Carb.Metab*, (*Dic./Mon.* vs. *C3/C4* only), it turns out to be significant at 5% probability for all countries, except the United States (**Figure 16**). This supplies initial evidence that C3 and C4 species with resistant biotypes to herbicides may not be equally distributed into dicots and monocots. In Brazil (**Figure 16**), most dicots are C3

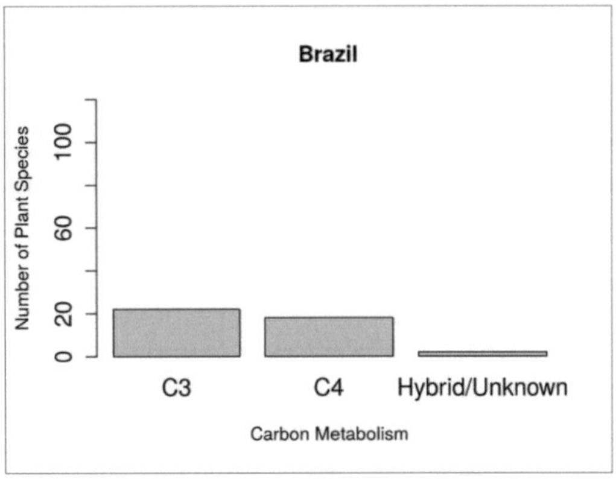


**Figure 15.** Occurrence of weed resistance as a function of carbon metabolism pathway in Brazil. Source: adapted from Heap [8].

while most monocots with resistance to herbicides are C4; for Australia and France, C3 species also predominate in the monocot class, while in Canada the proportion of C3 and C4 species with any biotype resistant to herbicides is equivalent (**Figure 16**).

In other words, it appears that the dicot class of angiosperms is significant (four out of four) for presenting a higher number of C3 species with resistant biotypes compared to C4. For monocots, Australia and France presented higher number of C3 species with resistance; in Canada, this relationship was alike, and in Brazil, there were more C4 monocot species with resistance to herbicides than C3. Anyway, the carbon metabolism pathway (**Figure 13** and **Figure 15**) seems to be the most significant compared to the angiosperm class (**Figure 12** and **Figure 14**). Thus, one would expect more cases of C3 weed biotypes with resistance to herbicides in Brazil (**Figure 15**), compared to the reference countries (**Figure 13**), for dicots (**Figure 16**). For monocots, there will be a need for a follow-up to understand if the tendency of majority in C4 species (**Figure 16**) will be maintained, or if it is only a deviation

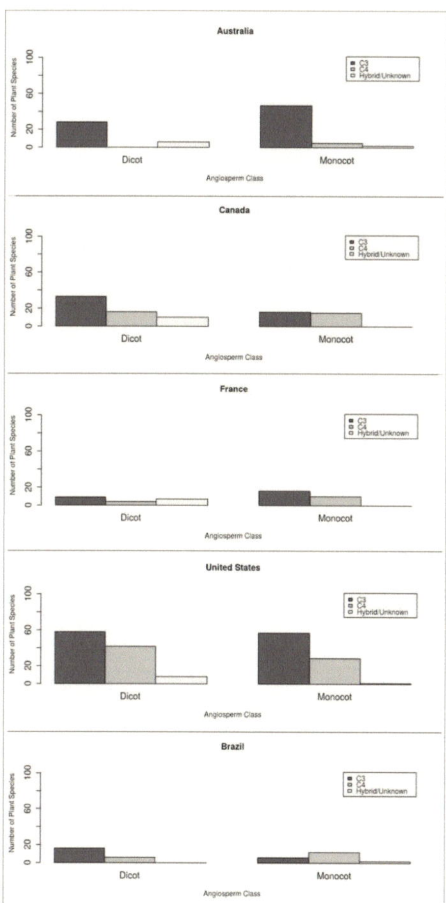


| | Dicots | Monocots | |
|---|---|---|---|
| **Country** | C3/C4 | C3/C4 | $X^2_{\text{p-value}}$ |
| Australia | ⇧ | ⇧ | 0.021 |
| Canada | ⇧ | ≅ | 0.018 |
| France | ⇧ | ⇧ | 0.004 |
| USA | ⇧ ns | ⇧ ns | 0.059 |
| Brazil | ⇧ | ⇩ | 0.014 |

Note: the $X^2$-test only supplies evidence if there is any association between two variables, or alternatively if they are independent. The interpretation presented in this table was subjectively obtained and should be used only as reference.

**Figure 16.** Occurrence of weed resistance as a function of botanical class and carbon metabolism pathway. Source: adapted from Heap [8].

from the real tendency which will be corrected by nature in the future. One should consider the probable loss of superiority from C4 plants over C3 as a consequence of the heavy anthropogenic effect in arable fields, as hypothesized by Ghannoum [24].

## 5. Geographical region of origin of families with resistant biotypes

The **probable** geographical center of origin of the families with resistant biotypes is summarized by the studied country in **Figure 17**. The region of origin for each botanical family is

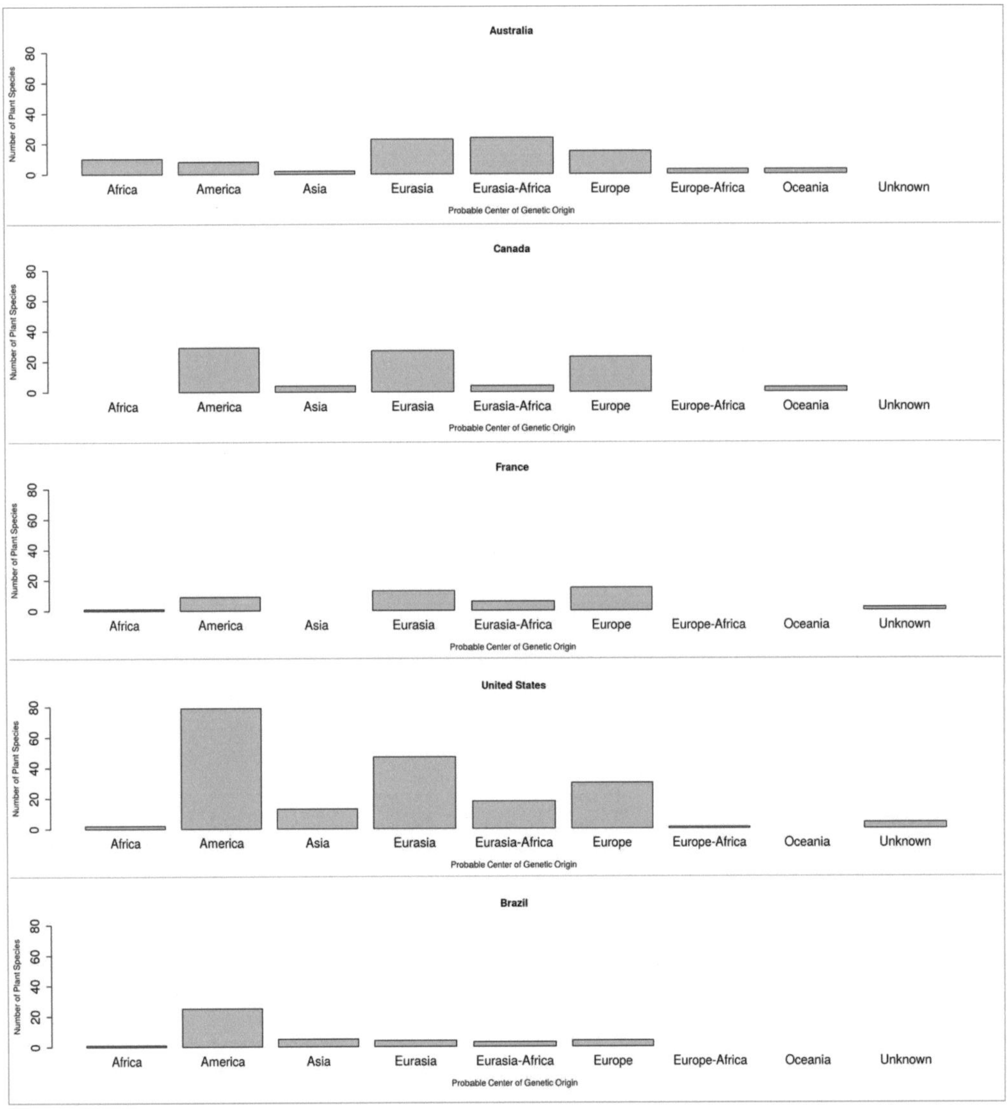


**Figure 17.** Occurrence of weed resistance as a function of the **probable** geographical region of origin of the genus. Source: adapted from Heap [8].

difficult to be defined, and data is sometimes controversial; thus, **Figure 17** should be interpreted as an approximation as close as possible to the currently available data about the origin of plant families. Families with higher degrees of uncertainties in their origin were grouped, like "Eurasia," which includes Europe and Asia, Eurasia-Africa (Europe, Asia, Africa), and Europe-Africa.

Most botanical families with resistant weed biotypes in Australia were originated in Eurasia and Africa; in Canada, they came from America and Eurasia (and of course Europe); in France, most weeds with resistant biotypes are native from Europe, and some of them could have come from Asia (Eurasia). In the United States, most families are native from the Americas, while about a half of the species with resistant biotypes came from Eurasia (in **Figure 17**, Asia + Eurasia + Europe data).

Summarizing, half or most of the weed species, which presented resistant biotypes in each of the reference countries, seem to be native to their continent (**Figure 17**), and this makes sense since the center of genetic origin of a given botanical group usually (if not always) presents the greatest genetic variation for that species [25]. Thus, the genetic variation which could result in the appearance and consequent selection of resistant biotypes would probably be most easily present in the genetic center of origin of the plant group. In Brazil (**Figure 17**), the same tendency is observed as most of the plants which presented resistant biotypes were most probably native from the Americas.

## 6. Most probable Brazilian weed groups to evolve resistance to herbicides

**Table 2** shows the main weed species in the Brazilian agriculture [2]. Surely, a great number of significant Brazilian weed species are out of the list, but the most cited ones in the specialized literature in soybean, corn, cotton, wheat, sugarcane, *Eucalyptus*, citrus, and cassava are included in **Table 2**. Orders and families with gray background are those identified in the reference countries as the most probable ones to contain weed species with superior ability to evolve resistance to herbicides. Genera and species with gray background are those that already present at least one resistant biotype in Brazil.

To date (November 2016), 20 weed species have been reported in Brazil as presenting at least one resistant biotype (**Table 2**). Three plant orders (Gentianales, Lamiales, and Solanales) are considered to contain weed species with superior ability to evolve resistance to herbicides, being these orders complemented by those that already present weed species with resistant biotypes (Asterales, Brassicales, Caryophyllales, Malpighiales, Alismatales, and Poales).

Eight botanical families are considered as presenting superior risk to evolve resistance to herbicides in Brazil (Asteraceae, Brassicaceae, Caryophyllaceae, Polygonaceae, Rubiaceae, Convolvulaceae, Solanaceae, and Poaceae), and five of them (Caryophyllaceae, Polygonaceae, Rubiaceae, Convolvulaceae, and Solanaceae) still do not present any weed species in Brazil with confirmed resistance to herbicides (**Table 2**).

Researchers should be aware, however, not to consider only the data summarized in **Table 2** to collect evidences about future cases of weed resistance in Brazil, as that table included

| Class | Order | Family | Genus / species | Common name | Common name (PT) | Main crops |
|---|---|---|---|---|---|---|
| Dicot | Apiales | Apiaceae | *Bowlesia incana* | Hoary bowlesia | Erva-salsa | Wheat |
| Dicot | Asterales | Asteraceae | *Acanthospermum hispidum* | Bristly starbur | Carrapicho-de-carneiro | Cotton, Citrus |
| Dicot | Asterales | Asteraceae | *Ageratum conyzoides* | Tropical whiteweed | Mentrasto | Cotton, Citrus |
| Dicot | Asterales | Asteraceae | *Bidens pilosa* | Hairy Beggarticks | Picão-preto | Cotton, Citrus, Cassava, Soybean, Wheat, Maize |
| Dicot | Asterales | Asteraceae | *Bidens subalternans* | Greater Beggarticks | Picão-preto | Cotton, Wheat |
| Dicot | Asterales | Asteraceae | *Conyza spp.* | Hairy Fleabane | Buva | Cotton, Eucaliptus, Cassava, Citrus, Soybean, Wheat |
| Dicot | Asterales | Asteraceae | *Emilia sonchifolia* | lilac tasselflower | Falsa-serralha | Citrus, Cassava |
| Dicot | Asterales | Asteraceae | *Galinsoga parviflora* | Gallant soldier | Picão-branco | Citrus, Wheat |
| Dicot | Asterales | Asteraceae | *Parthenium hysterophorus* | Ragweed parthenium | Losna-branca | |
| Dicot | Asterales | Asteraceae | *Sonchus oleraceus* | Sow thistle | Serralha | Citrus, Wheat |
| Dicot | Asterales | Asteraceae | *Synedrellopsis grisebachii* | Straggler daisy | Agriãozinho | Citrus, Eucaliptus |
| Dicot | Asterales | Asteraceae | *Tridax procumbens* | Coatbuttons | Erva-de-touro | Citrus, Eucaliptus |
| Dicot | Brassicales | Brassicaceae | *Raphanus* spp. | | | Cassava, Wheat, Maize |
| Dicot | Caryophyllales | Amaranthaceae | *Alternanthera tenella* | Parrotleaf | Apaga-fogo | Cotton |
| Dicot | Caryophyllales | Amaranthaceae | *Amaranthus spp* | | Caruru | Cotton, Sugarcane, Citrus, Maize, Cassava |
| Dicot | Caryophyllales | Caryophyllaceae | *Stellaria media* | Common chickweed | Erva-de-passarinho | Wheat |
| Dicot | Caryophyllales | Polygonaceae | *Rumex* spp | | Língua-de-vaca | Wheat |

http://dx.doi.org/10.5772/intechopen.68336

| Class | Order | Family | Genus / species | Common name | Common name (PT) | Main crops |
|---|---|---|---|---|---|---|
| Dicot | Caryophyllales | Portul acaceae | *Portulaca oleracea* | | Beldroega | Cassava, Sugarcane, Citrus |
| Dicot | Cucurbitales | Cucurbitaceae | *Luffa aegyptiaca* | Spongegourd | Bucha | Sugarcane |
| Dicot | Cucurbitales | Cucurbitaceae | *Momordica charantia* | Bitter melon | Melão-de-são-caetano | Sugarcane |
| Dicot | Fabales | Fabaceae | *Aeschynomene spp.* | | Angi quinho | Rice |
| Dicot | Gentianales | Rubiaceae | *Borreria verticillata* | Buttonweed | Vassoura-de-botão | Cotton |
| Dicot | Gentianales | Rubiaceae | *Richardia brasiliensis* | Brazilian calla-lily | Poaia-branca | Citrus, Maize, Wheat |
| Dicot | Gentianales | Rubiaceae | *Spermacoce latifolia* | Malayalam | Erva-quente | Citrus, Eucaliptus |
| Dicot | Lamiales | Boraginaceae | *Echium plantagineum* | Patersons curse | Flor roxa | Wheat |
| Dicot | Malpighiales | Euphorbiaceae | *Euphorbia heterophylla* | Wild Poinsettia | Leiteiro | Cotton, Sugarcane, Cassava, Maize, Soybean, Wheat |
| Dicot | Malvales | Malvaceae | *Sida* spp. | Sida | Guanxuma | Cassava, Wheat, Sugarcane, Maize |
| Dicot | Myrtales | Onagraceae | *Ludwigia longifolia* | Primrose willow | Cruz-de-malta | Rice |
| Dicot | Solanales | Convolvulaceae | *Ipomoea spp.* | Morningglory | Corda-de-viola | Cotton, Citrus, Cassava, Soybean, Wheat, Maize |
| Dicot | Solanales | Convolvulaceae | *Merremia aegyptia* | Hairy merremia | Corda-de-viola | Sugarcane |
| Dicot | Solanales | Convolvulaceae | *Merremia cissoides* | Roadside woodrose | Corda-de-viola | Sugarcane |
| Dicot | Solanales | Solanaceae | *Nicandra physalodes* | Apple-of-Peru | Joá-de-capote | Cotton |
| Dicot | Solanales | Solanaceae | *Solanum americanum* | American black nightshade | Maria-pretinha | Cotton |
| Dicot | Solanales | Solanaceae | *Solanum viarum* | Tropical soda apple | Joá-bravo | Cotton |

| Class | Order | Family | Genus / species | Common name | Common name (PT) | Main crops |
|---|---|---|---|---|---|---|
| Monocot | Alismatales | Alismataceae | *Sagittaria montevidensis* | Giant arrowhead | Chapéu-de-couro | Rice |
| Monocot | Commelinales | Commelinaceae | *Commelina benghalensis* | Benghal dayflower | Trapoeraba | Cotton, Eucaliptus, Cassava |
| Monocot | Commelinales | Pontederiaceae | *Heteranthera reniformis* | Kidney leaf mud plantain | Aguapé | Rice |
| Monocot | Poales | Cyperaceae | *Cyperns spp.* | Sedges | Ciperáceas | Cassava, Sugarcane, Citrus, Rice |
| Monocot | Poales | Cyperaceae | *Fimbristyllis miliacea* | Fringerush | Cuminho | Rice |
| Monocot | Poales | Poaceae | *Avena sativa* | Wild oat | Aveia | Wheat |
| Monocot | Poales | Poaceae | *Avena strigosa* | Wild oat | Aveia | Wheat |
| | | | *Avena fatua* | Wild oat | Aveia | |
| Monocot | Poales | Poaceae | *Brachiaria spp.* | Alexandergrass | Capim-marmelada | Sugarcane, Citrus, Cassava, Maize, Soybean, Wheat, Rice |
| Monocot | Poales | Poaceae | *Cenchrus echinatus* | Southern sandbur | Capim-carrapicho | Cotton, Citrus, Cassava |
| | | | *Chloris elata* | Tall windmill grass | Capim-branco | |
| Monocot | Poales | Poaceae | *Cynodon dactylon* | Vilfa stellata | Grama-seda | Sugarcane, Citrus |
| Monocot | Poales | Poaceae | *Digitayia spp.* | Sourgrass | Capim amargoso | Citrus, Eucaliptus, Cassava, Soybean, Cotton, Maize, Rice |
| Monocot | Poales | Poaceae | *Echinochha spp* | Barnyardgrass | Capim arroz | Maize, Rice |
| Monocot | Poales | Poaceae | *Eleusine indica* | Goosegrass | Capim pé-de-galinha | Cotton, Sugarcane, Citrus, Maize, Rice |
| Monocot | Poales | Poaceae | *Eriochloa punctata* | Louisiana cupgrass | Capim-de-várzea | Rice |

| Class | Order | Family | Genus / species | Common name | Common name (PT) | Main crops |
|---|---|---|---|---|---|---|
| Monocot | Poales | Poaceae | *Ischaemum rugosum* | Ribbed muraingrass | Capim-macho | Rice |
| Monocot | Poales | Poaceae | *Leersia hexandra* | Southern cutgrass | Grama-boiadeira | Rice |
| Monocot | Poales | Poaceae | *Lolium multiflorum* | Italian ryegrass | Azevém | Eucaliptus, Maize, Soybean, Wheat |
| Monocot | Poales | Poaceae | *Luziola peruviana* | Peruvian watergrass | Grama-boiadeira | Rice |
| Monocot | Poales | Poaceae | *Oryza sativa* | Weedy rice | Arroz daninho | Rice |
| Monocot | Poales | Poaceae | *Panicum dichotomiflorum* | Fall panicgrass | Capim-do-banhado | Rice |
| Monocot | Poales | Poaceae | *Panicum maximum* | Guinea grass | Capim-colonião | Sugarcane, Citrus, Cassava |
| Monocot | Poales | Poaceae | *Paspalum modestum* | Water paspalum | Lombo-branco | Rice |
| Monocot | Poales | Poaceae | *Rotboellia exaltata* | Itchgrass | Capim-camalote | Sugarcane |
| Monocot | Poales | Poaceae | *Sorghum halepense* | Johnson grass | Massambará | Sugarcane |

NOTE: Botanical orders and families marked with gray background present superior potential risk of including weed species with resistant biotypes in the future based on the history of the reference countries (Australia, Canada, France and United States). Genus/Species with gray background already include weed species with resistant biotypes in Brazil as of November, 2016.

**Table 2.** Botanical classification highlighting orders, families and Genus/species of weeds with superior ability to evolve resistance in Brazil.

only a few weed species from the total Brazilian pool of weed species listed by some authors [3, 26–28]. The additional data supplied in the present chapter (herbicide mechanism of action, carbon metabolism, geographical region of origin, etc.) should be also considered together with the list of herbicides available for each crop grown in Brazil, as well as the frequency of application of each herbicide in each crop.

## 7. Conclusions

Most weed species with resistant biotypes in the reference countries seem to be native to their continent. The most important botanical families with resistant biotypes in the reference countries were also among the first ones to appear in the respective countries. There was predominance of C3 species over C4 in the number of plant species with resistant biotypes in the reference countries. In Brazil, three orders (Gentianales, Lamiales, and Solanales) are considered as high risk, besides the six already present. Furthermore, eight botanical families present superior risk to evolve resistance, and for five of them (Caryophyllaceae, Polygonaceae, Rubiaceae, Convolvulaceae and Solanaceae) resistance cases have not been reported to date in Brazil.

## Author details

André Andres[1], Germani Concenço[1*], Fábio Schreiber[2], Dirceu Agostinetto[2], Leandro Vargas[3], João Behenck[2], Giovanni Antoniaci Caputo[2] and Ygor Sulzbach Alves[2]

*Address all correspondence to: germani.concenco@embrapa.br

1 Embrapa Temperate Agriculture, Pelotas, RS, Brazil

2 Federal University of Pelotas, RS, Brazil

3 Embrapa Wheat, Passo Fundo, RS, Brazil

## References

[1] Silva AA, Silva JF. Topics in Weed Management. Viçosa: UFV; 2007. p. 367

[2] Monquero PA. Weed Management in Agricultural Crops. Sao Carlos: RiMa; 2014. p. 288

[3] Lorenzi H. Brazilian Weeds. Nova Odessa: Instituto Plantarum; 2008. p. 640

[4] Andres A, Fogliatto S, Ferrero A, Vidotto F. Growth variability of Italian weedy rice populations grown with or without cultivated rice. Crop Science. 2015;**55**:394-400

[5] Aldrich RJ. Weed-Crop Ecology: Principles in Weed Management. North Scituate: Breton; 1984. p. 465

[6] Agostinetto D, Vargas L. Weed Resistance to Herbicides in Brazil. Pelotas: UFPel; 2014. p. 398

[7] Concenço G. Evolution, epigenetics, resistance: Troublesome weeds. Revista Brasileira de Herbicidas. 2016;**15**(1):14-25

[8] Heap I. The International Survey of Herbicide Resistant Weeds [Internet]. 2016. Available from: www.weedscience.org

[9] Jomini P, Boulanger P, Zhang X, Costa C, Osborne M. The Common Agricultural Policy and The French, EU and Global Economies. Paris: Groupe d'Économie Mondiale; 2009. p. 38

[10] DG Agriculture and Rural Development. Analysis of the EU Fruit and Vegetable Sector. Brussels: European Union; 2014. p. 40

[11] MAPA – Ministry of Agriculture, Livestock and Distribution. Projections of Agribusiness. MAPA, Brazil. 2015. Available from: http://www.agricultura.gov.br

[12] Dahr HE. Agriculture, a Strategic Sector for Brazil's Economic Growth [Internet]. 2012. Available from: http://agriculture.gouv.fr/Analyse-no41-mars-2012-Entre

[13] Moreira JR, Medeiros MB. The Legacy of Darwin and the Agricultural Research. Brasilia: Embrapa; 2014. p. 341

[14] Souza VC, Lorenzi H. Systematic Botany. Nova Odessa: Instituto Plantarum; 2008. p. 704

[15] Christoffoleti PJ, Nicolai M. Aspects of the Weed Resistance to Herbicides. 4th ed. Piracicaba: ESALQ; 2016. p. 262

[16] Christoffoleti PJ. Aspects of the Weed Resistance to Herbicides. 3rd ed. Piracicaba: ESALQ; 2008. p. 120

[17] Spooner D, Hetterscheid W, Van Der Berg R, Brandenburg W. Plant nomenclature and taxonomy. Horticultural Reviews. 2003;**28**:1-60

[18] Dudzik M. Seed plants: Gymnosperms and angiosperms. Connexions Module: m48092. 2013. Available from: http://cnx.org/content/m48092/1.1/

[19] Barkley TM. History of plant taxonomy. In: Radford AE, editor. Fundamentals of Plant Systematics. New York: Harper & Row; 1986. pp. 37-59

[20] Cronquist A. An Integrated System of Classification of Flowering Plants. New York: Columbia University Press; 1981. ISBN 9780231038805

[21] Cronquist A. The Evolution and Classification of Flowering Plants. 2nd ed. Bronx, NY: New York Botanical Garden; 1988 [1968]

[22] Gurevitch J, Scheiner SM, Fox GA. Plant Ecology. Porto Alegre: Artmed; 2009. p. 574

[23] Ehleringer JR, Monson RK. Evolutionary and ecological aspects of photosynthetic pathway variation. Annual Review of Ecology and Systematics. 1993;**24**:411-439

[24] Ghannoum O. C4 photosynthesis and water stress. Annals of Botany. 2009;**103**(4):635-644

[25] Barbieri RL, Stumpf ERT. Origin and Evolution of Cultivated Plants. Brasilia: Embrapa; 2008. p. 914

[26] Kissmann KG. Infesting and Noxious Weeds. Vol. 1. São Paulo: Basf; 2000. p. 825

[27] Kissmann KG, Groth D. Infesting and Noxious Weeds. Vol. 2. São Paulo: Basf; 2000. p. 978

[28] Kissmann KG, Groth D. Infesting and Noxious Weeds. Vol. 3. São Paulo: Basf; 2000. p. 726

Chapter 7

# Procedures for Detection of Resistant Weeds Using $^{14}C$-Herbicide Absorption, Translocation, and Metabolism

Kassio Ferreira Mendes, Ricardo Ferraz Silveira,
Miriam Hiroko Inoue and Valdemar Luiz Tornisielo

Additional information is available at the end of the chapter

http://dx.doi.org/10.5772/68092

**Abstract**

Herbicide resistance mechanisms involve altered absorption, translocation, and metabolism of herbicides (i.e., glyphosate), and this is an important component in the study of herbicide resistance mechanisms as well. $^{14}C$-herbicides are used in resistant weeds studies, since they provide some advantages in comparison with chemical measures, including greater sensitivity, stepwise description of a particular element in a metabolic system, herbicide position, detection through X-ray films and/or radio image, and liquid scintillation. However, an up-to-date, organized description and standardization of research procedures and methodology on the use of radioisotopes for detection of resistant weeds, through different mechanisms of absorption, translocation, and metabolism in comparison with susceptible weeds are lacking in the literature. Techniques that use $^{14}C$ such as tracers are extremely useful to study the herbicides behavior in the resistant weed, since the radiometric techniques offer the possibility of accurately determining very small amounts in a relatively short time. However, mechanism of resistance to herbicides in this resistant weed population compared with the susceptible population cannot be due to differential absorption, translocation, or metabolism of herbicide in weed; so other studies are necessary to elucidate the mechanism of herbicide resistance on weed population.

**Keywords:** mechanism of resistance, metabolites, standard methodology, radioisotopes

## 1. Introduction

Herbicides can penetrate plants through their aerial structures (leaves and stems), subterraneous (root, rhizome, stolon, and tuber), and young structures such as radicles and caulicles. The main route of penetration of the herbicides in the plant is a function of a series of intrinsic and extrinsic (environmental) factors. Absorption of herbicides by roots or leaves is influenced by the availability of the products at the sites of absorption and environmental factors (temperature, light, relative humidity, and soil moisture), which also influences the translocation of these to the site of action [1].

Among the biochemical and physiological mechanisms, the change in the absorption, translocation, or metabolism of resistant weed biotypes has been reported on several species for different herbicides. These resistance mechanisms have been studied over the last years, allowing the development and improvement of analytical techniques to diagnose this type of resistance [2]. However, an up-to-date, organized description and standardization of research procedures and methodology on the use of radioisotopes for detection of resistant weeds, through different mechanisms of absorption, translocation, and metabolism in comparison with susceptible weeds are lacking in the literature.

Radioisotopes are used on several research areas, such as for the metabolism of drugs and pesticides, environmental studies to determine biological routes and mass balance studies for organic compounds, and the ones that are most frequently used are tritium and $^{14}C$. The method for using radiolabeled herbicides may be quantitative or qualitative, allowing associating the resistance to the reduced absorption and/or translocation, and/or to the accelerated metabolism in several weed species [3]. Therefore, it is important to understand concepts and measurement units of the main analytical techniques that use labeled molecules with $^{14}C$ to study the biochemical and physiological resistance mechanisms to herbicides, as well as for studies that evaluate the destination of these molecules on the environment. Understanding these mechanisms is fundamental for management alternatives to be planned or to improve the effectiveness of the product [4].

Considering the above, the objective of this chapter was to conduct a description of the research procedures and the methodology related for detection of resistant weeds using $^{14}C$-herbicide absorption, translocation, and metabolism compared with susceptible weeds.

## 2. Restriction of herbicide movement in resistant weeds

As long as a plant biotype is susceptible to an herbicide, the biological activity resulting from the pulverization of the herbicide in the plant is dependant of the absorption and translocation of that herbicide in the plant.

Translocation is a desirable attribute because it allows the herbicide to reach both treated and untreated parts of the plant [5]. It is especially important when used for controlling plants that are able to regenerate themselves through structures such as bulbs, rhizomes, stolons, and tubers. If, for some reason, the herbicide fails to reach these structures due to restriction of movement, the plants are not going to be controlled and will therefore be resistant.

Weed Science Society of America (WSSA) defines herbicide resistance as the inheritable ability of a plant biotype to survive and reproduce following exposure to an herbicide dose that would normally be lethal to the wild type [2].

Resistance conferred by the restriction of herbicide movement mechanism is classified as non-target-site resistance (NTSR). Weeds that are resistant due to this mechanism commonly show higher foliar retention and reduced translocation, reducing the amount of herbicide that reaches the target, making it insufficient to exercise control over the weed.

Goggin et al. [6] employed $^{14}C$-labeled 2,4-dichlorophenoxyacetic acid (2,4-D) to study resistance in two wild radish (*Raphanus raphanistrum* L.) biotypes from Australia. When comparing with a susceptible population, results showed that the resistance is due to an inability to translocate 2,4-D out of the treated leaf. Further investigation is necessary, but the authors suggest that the restriction of herbicide movement could be due to an alteration in the activity of a plasma membrane ABCB-type auxin transporter responsible for facilitating long-distance transport of 2,4-D.

Reduced translocation was reported as the cause of resistance to paraquat in two populations of *Hordeum leporinum* [7]. The inability to translocate paraquat out of the treated leaves was verified with the use of $^{14}C$-labeled paraquat comparing the two resistant populations with a susceptible one, all from Australia.

Riar et al. [8] studied three barnyardgrass (*Echinochloa crus-galli*) biotypes from the United States with cross-resistance to imazamox, imazethapyr, penoxsulam, and bispyribac-sodium. The authors concluded that reduced translocation could contribute to imazamox and bispyribac-sodium resistance for two out of three biotypes.

Regarding glyphosate, the world's most important and widely used herbicide, NTSR has been reported as one of the most widespread type of resistance [9].

Glyphosate is a foliar applied herbicide which follows a source-to-sink pattern and kills plants through the inhibition of 5-enolpyruvylshikimate-3-phosphate synthase (EPSPS), which are most highly expressed in the meristems and flowers of plants [10]. Since it is applied on the shoots, it must traverse the non-living structures of the leaf cuticle and the cell walls of the epidermis, apoplast, and mesophyll prior to accessing the phloem for transport to sink tissues [11]. Glyphosate's great ability to translocation in the plant reaching vital areas such as the roots and shoot meristems is one of the characteristics that makes it so important and efficient, but it also makes it highly dependent on herbicide movement.

Ferreira et al. [12] reported an increase in foliar retention in hairy fleabane (*Conyza bonariensis*) resistant to glyphosate. Reduced translocation was reported to be one of the mechanism conferring resistance to glyphosate in rigid ryegrass (*Lolium rigidum*) [13] and perennial ryegrass (*Lolium perenne*) [14].

The mechanism of glyphosate absorption into plant cells is not well understood. There appears to be two different mechanisms of absorption. One is an active system that pumps the herbicide into plant cells, possibly via a phosphate transporter, and operates at low concentrations. Other may be a passive mass flow system which is gradient dependent (**Figure 1**).

The exact mechanism that promotes the reduction of cellular absorption and translocation of glyphosate in resistant weeds is not clear yet. Shaner [10] described four potential mechanisms that may cause the restriction of glyphosate movement (**Figure 2**): (1) alteration in a putative phosphate transporter responsible for the active cellular absorption of glyphosate, in a way that the transporter is no longer present or no longer recognizes glyphosate, resulting in reduced absorption and translocation; (2) evolution of a new transporter that pumps glyphosate into the vacuole, thus sequestering the herbicide and preventing it from reaching either the chloroplast or the phloem; (3) evolution of a new transporter that actively pumps glyphosate out of the cell into the apoplast; or (4) evolution of a transporter at the chloroplast envelope that pumps glyphosate out of the chloroplast, preventing the herbicide from reaching its target site.

In order to study glyphosate resistance, $^{31}$P nuclear magnetic resonance (NMR) spectroscopy studies were employed to track glyphosate movement and metabolism in resistant and susceptible biotypes of horseweed (*Conyza canadensis*), and the results showed that the rate of vacuole accumulation of this herbicide is faster and occurs to a greater extent in the resistant biotype rather than in the susceptible [15].

These results have been confirmed in different glyphosate-resistant *Lolium* spp. biotypes collected on three different continents [16], pointing to vacuolar glyphosate sequestration as the primary mechanism of resistance in these biotypes.

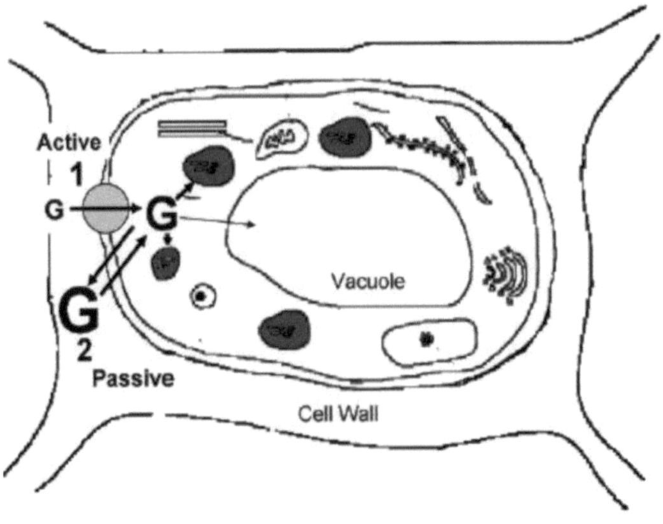


**Figure 1.** Proposed mechanisms of glyphosate absorption into plant cells. G, glyphosate (the size of the letter indicates relative size of glyphosate pool). (1) Active absorption of glyphosate into cell. (2) Passive diffusion of glyphosate into the cell. Arrows indicate direction of movement of glyphosate pools into and out of the cell, chloroplast, and vacuole. Source: Shaner [10].

http://dx.doi.org/10.5772/68092

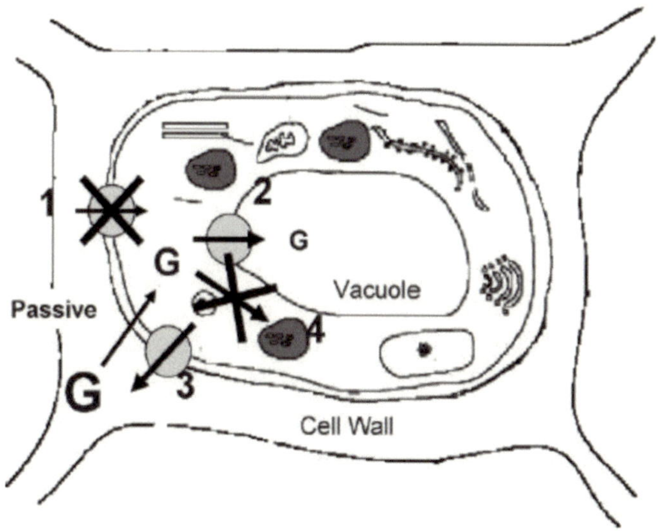


**Figure 2.** Potential mechanisms for reduced glyphosate cellular absorption in glyphosate-resistant (GR) biotypes. (1) Inhibition of active absorption by a modification of active transporter. (2) An active transporter that pumps glyphosate into the vacuole. (3) An active transporter that pumps glyphosate from the cell into the apoplast. (4) Inhibition of glyphosate absorption into the chloroplast by a transporter that pumps it out of the chloroplast. G, glyphosate. Source: Shaner [10].

# 3. Preparation of resistant weed samples by oxidizer

According IRSN [17], the $^{14}C$ contained in the resistant weed (test portion) is transformed to $^{14}CO_2$ from which a sample is prepared for measurement by liquid scintillation spectrometry (LSS), and combustion by oxidizer (**Figure 3**) is main method used.

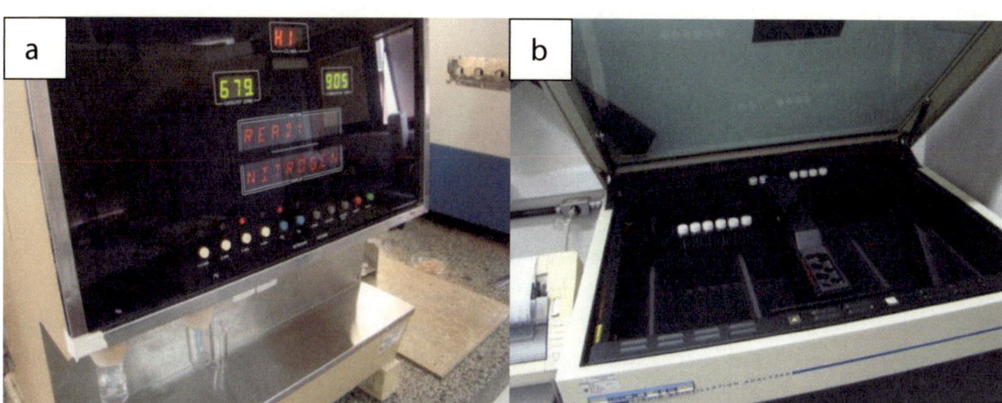


**Figure 3.** Oxidizer OX500 (R.J. Harvey Instrument Corporation) (a) and liquid scintillation equipment, Tri-Carb 2910 TR LSA counter (PerkinElmer) (b) from the Laboratory of Ecotoxicology of CENA/USP.

Resistant weed samples are not readily soluble on scintillation cocktails. Due to this reason, such samples go through biological combustion on oxidizer. The combustion of the sample creates an atmosphere that is rich in hydrogen, which is oxidized by the water, while the entire carbon content is oxidized by the carbon dioxide containing $^{14}C$ ($^{14}CO_2$). Evolved $^{14}CO_2$ is trapped in a 2 M NaOH solution and subsequently mixed in an adequate scintillating cocktail for ß counting on a LSS [18].

Coughtrey et al. [19] described a wet oxidation technique using potassium dichromate and concentrated sulfuric and phosphoric acid, which can be done in a modified filter flask. This technique can accommodate up to 0.3 g of dry resistant weed. Recovery of $^{14}C$ is consistent between batches, with an average recovery of 97.2% over 15 standards. These authors reported that technique described does not involve large capital expenditure and is relatively rapid.

The expression of the resistant weed sample's activity in becquerel (Bq) of $^{14}C$ per kg of carbon also requires measuring its elementary carbon content, generally by gas chromatography. According Nandula and Vencil [20], the commonly accepted unit of measurement of radioactivity is the Bq, derived from the International System of Units. It is defined as follows:

$$1 \text{ becquerel } (\text{Bq}) = 1 \text{ disintegration/s } (\text{dps}) = 60 \text{ disintegrations/min } (\text{dpm}) \quad (1)$$

A description of the research procedures and the methodology related for detection of resistant weeds using $^{14}C$-herbicide absorption, translocation, and metabolism compared with susceptible weeds will be described below, based on Nandula and Vencil [20] and Mendes et al. [21].

## 4. Herbicide absorption and translocation in resistant weeds

Studies on the absorption and translocation of herbicides in plants are usually conducted to evaluate the behavior of a new herbicide on a certain plant species, comparing two or more herbicides, specific formulations, additives, or the effect of environmental standards. The growing problem regarding the resistance of weeds to herbicides promoted the studies on the absorption, translocation, and metabolism of herbicides as the methodology to elucidate the resistance mechanisms [20]. So these procedures need to be better explained to researchers, as will be described in this chapter.

The studies on the absorption of herbicides use a destructive sampling of treated plants on several post-treatment periods, which allows the characterization of the absorption standard on the plant, considering the planning and adequate statistical analyses [22].

On the adequate phenological stage for each species, susceptible and resistant weeds must be adequately identified by treatment. The leaves that have been predetermined to receive the radiolabeled herbicide must be covered with plastic film, aluminum paper, or small paper envelopes. Then, the "cold" herbicide is applied to the plants (without the radioisotope) at the dose recommended by the manufacturer, as a solution with adjuvant (when indicated) and water, followed by the immediate removal of the protective plastic film of the applied leaf.

The radiolabeled herbicide solution must be prepared on a solution containing its commercial formulation at the recommended dose for the considered phenological stage. After applying

http://dx.doi.org/10.5772/68092

the "cold" product, its radiolabeled version is applied. It is important for the radiolabeled herbicide to be applied with at least 170 Bq of specific activity, in the case of studies with most of the annual weeds [20].

The radiolabeled product is applied using a micro-syringe, by applying a 1 μL droplet (the total radiolabeled product applied depends on the molecule and the radioactivity of the radiolabeled molecule), on the leaf blade of the upper part of the expanded leaf of each plant (**Figure 4**). The choice for the leaf on which the application will occur depends on the studied species. Each plant (or part of the plant) must be collected according to the pre-established times for each situation. However, it is suggested that at least six collection times are used, in addition to time zero (immediately after the application), and that the untreated plants are included as control. For each collection, the treated leaf from each plant must be rinsed with the adequate solvent. The concentration ($v\ v^{-1}$) of the solvent must be established on preliminary tests with the studied molecule. Then, the radioactivity during the rinsing must be quantified by LSS in order to determine the non-absorbed radioactivity. The leaf absorption is calculated by the difference between the applied and the non-absorbed radioactivity. The plants must be dried with an absorbing paper, pressed, and dried on an air circulation oven at 70°C for 48 h.

In the preparation of the absorption studies, we must select resistant and susceptible weeds of the same age and/or growth stage. According Nandula and Vencill [20] to plot, the figure is necessary use at least six time points in addition to a 0 time point of tissue harvest, as illustrated in **Figure 5**. However, under conditions of limited resources, it is better to increase the number of time points and reduce the number of replications ($n \geq 2$). Include non-treated weeds as a blank or control is very important for research. Then, the steps to evaluate the translocation are conducted.

**Figure 4.** Application of $^{14}C$-glyphosate with a micro-syringe on glyphosate tolerant *Spermacoce verticillata* leaves at the Laboratory of Ecotoxicology of CENA/USP.

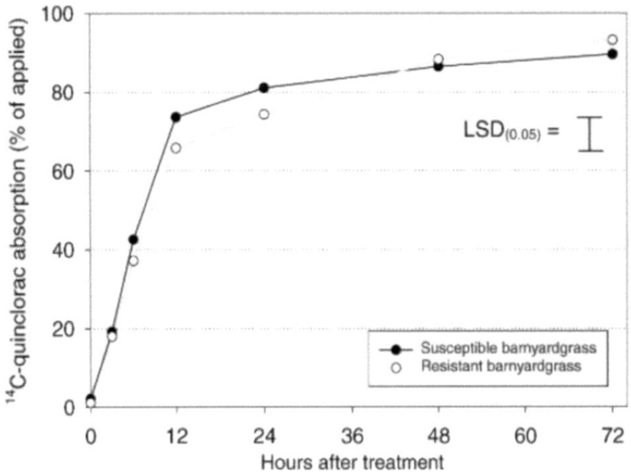


**Figure 5.** Absorption of $^{14}C$-quinclorac by propanil- and quinclorac-resistant and susceptible barnyardgrass (*Echinochloa crus-galli*) biotypes over time. No differences were detected between biotypes at any time. LSD (0.05) bar to make comparisons between biotypes at a particular time interval. Source: Lovelace et al. [23].

Usually, the translocation studies are conducted right after the absorption studies, although they demand more work and time. Differently from the absorption, which occurs within hours after the treatment, the translocation of herbicides may take up to days after the treatment. Due to this reason, in order to evaluate the translocation, the previous knowledge must be considered in order to determine the times after the treatment in which this variable should be evaluated.

The biological combustion is the most used procedure to quantify the translocation of herbicides on plants. However, care must be taken when stating that the detection of the radioactivity on other parts of the plant, outside the treated leaf, means that the herbicide is on its parental form. It might have been converted into a non-phytotoxic metabolite. In order to state this, one must investigate the potential for the herbicide to have been metabolized by the studied weed, through the information available in the literature.

To study the movement of herbicides on weeds, the qualitative techniques involving autoradiography or phosphorus blade images have been used for over 50 years [20]. While the biological combustion offers a quantitative estimation of the herbicide on the treated weed, autoradiography (**Figure 6**), or the phosphorus blade image provides a qualitative measurement of the movement of the herbicide on the weed, in addition to the location where it occurs.

For the exposition of the treated and untreated plants, the use of phosphorus blade images is safer in comparison to the use of autoradiography, since it does not require handling chemical compounds that are harmful to the health. Despite more expensive, the technique is also quicker. A single day of exposition of a plant on a phosphorus blade resulted on images with superior quality than the exposition for 3 weeks with the X-ray film [24].

Therefore, in order to study the translocation, the plants treated as on the absorption study must be exposed on phosphorus blade for 72 h, in order to scan the image for qualitative

http://dx.doi.org/10.5772/68092

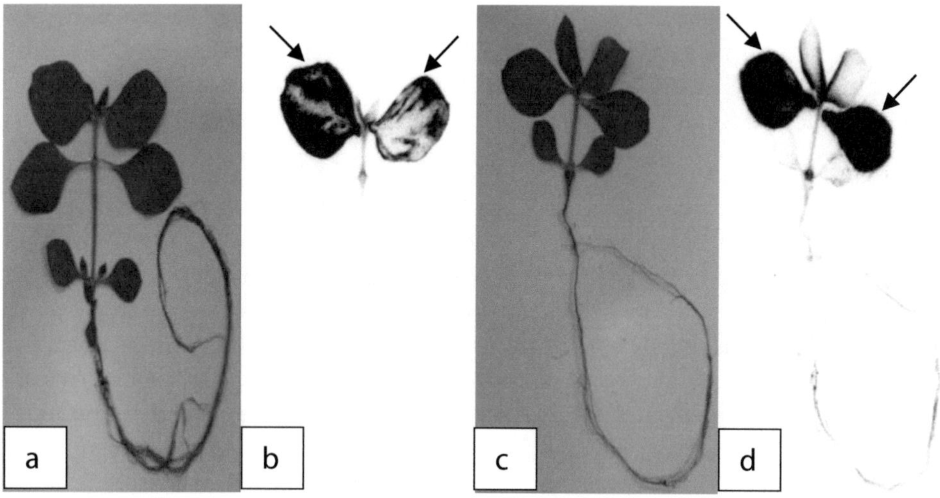


**Figure 6.** Autoradiography of glyphosate tolerant (a and b) and susceptible (c and d) *Richardia brasiliensis* with application of the leaf of $^{14}C$-glyphosate at 48 HAT (hours after the treatment). Photograph weed to the right (a and c) and autoradiography of the weed translocation to the left (b and d) at the Laboratory of Ecotoxicology of CENA/USP. Arrows indicate the sites of application.

analysis. The usual procedure to quantify the translocation of herbicides on plants is the biological combustion, in which dry samples of each part of the plant (both the treated leaf and the part above and below it, as well as the roots) are oxidized by the presence of $O_2$, and the resulting $CO_2$ is captured on a special solvent. Then, the radioactivity must be measured on the scintillation counter.

The quantitative analysis of the translocation may also be conducted through the volume analysis, offered by the software provided together with the image scanner, as of its purchase. The volume is the total signal intensity of the radioactivity within defined limits of the image. The translocation is then expressed as the rate between the percentage of signal intensity on the applied zone, as well as above and below it, and the total signal intensity on a defined image containing $^{14}C$ [25].

## 5. Herbicide metabolism in resistant weeds

The use of radiolabeled herbicides to investigate whether the herbicide is being metabolized in the resistant weed is an efficient method, and it is the most indicated method to diagnose the resistance related to other phenomena that are not related to the change on the action site of the herbicide [26]. The analytical method aiming at studying the metabolism of herbicides in plants comprehends three fundamental steps: preparation of the plants and application of treatments; extraction and separation; and identification of the herbicide and its metabolites, if any.

The steps to conduct the study on the metabolism of herbicides in plants are described as follows. The preparation of plants and application of the treatments must be conducted as described for the absorption and translocation study. In case the fresh samples of plants are not adequate for processing after the collection, the ideal is to store them at −20°C to assure the stability of the active substances and metabolites. The techniques employed on studies on the metabolism of herbicides in plants are thin layer chromatography (TLC), high-performance liquid chromatography (HPLC), and gas chromatography (GC), depending on the herbicide molecule.

For the extraction, the adequate system of solvents for the studies herbicide must be known. The treated leaf must be rinsed with non-polar solvent (usually ethanol or methanol). Then, the plant must be dried with an absorbing paper, immediately frozen in liquid nitrogen and stored at −80°C up to its use. The plant tissue must be macerated in crucibles that must be previously cooled with $N_2$, and homogenized with the specific cold solvent at a concentration of 80% (v $v^{-1}$). A stainless steel homogenizer may also be used. The solution must be centrifuged; the supernatant decanted; and the residue must go through re-extraction with the chosen cold solvent at 80%, followed by extraction with the same cold solvent at 50% (v $v^{-1}$). The supernatants must be mixed, and the radioactivity must be determined by LSS, in order to know the mass balance, which is expressed as the rate between the radioactivity applied at the beginning of the experiment and the total radioactivity measured (originated from rinsing all parts of the plant). The mass balance may be also referred to as the radioactivity recovery percentage. Approximately 7 mL of the supernatant must be evaporated, resuspended in 300 mL of the solvent at 50%, and centrifuged. The final sample may be analyzed by any previously described technique, usually TLC or HPLC, with the respective solvent system [4].

## 6. Results of differential absorption, translocation, and metabolism of herbicides in resistant and susceptible weeds

Several researchers have studied herbicides behavior in weeds in order to find resistance mechanics through the differential of $^{14}$C-herbicide absorption, translocation, and metabolism, according to **Table 1**. These results suggest that reduced translocation and accelerated metabolism of herbicide plays a major role in herbicide resistance in resistant biotypes of weed. Likewise, differences in absorption may contribute to the differential sensitivity of herbicide resistant and susceptible weed populations.

Overall, herbicide absorption is similarly compared with resistant and susceptible biotypes, but the difference in herbicide translocation is notorious in most studies reported (**Table 1**). Although differential translocation can be observed between resistant and susceptible weeds, it is unclear whether this difference is a cause of herbicide resistance or an effect of some other physiological process [23].

Herbicide metabolism studies are not always researched together with herbicide absorption and translocation studies, because the increased herbicide metabolism in resistant biotypes compared with susceptible transforms this product on metabolites without herbicidal action (**Table 1**). Among herbicides reported, glyphosate is more studied. Studies on glyphosate metabolism, expression, and sensitivity of target enzyme EPSPS synthase

| Herbicide | Weed | Biotype | Absorption | Translocation | Metabolism | Reference |
|---|---|---|---|---|---|---|
| Clopyralid | *Centaurea solstitialis* | Resistant | = | = | ↑ | Valenzuela et al. [27] |
| | | Susceptible | = | = | ↓ | |
| Quinclorac | *Echinochloa crus-galli* | Resistant | = | ↓ | NA | Lovelace et al. [23] |
| | | Susceptible | = | ↑ | NA | |
| Glyphosate | *Lolium multiflorum* | Resistant | = | ↓ | NA | Pérez et al. [28] |
| | | Susceptible | = | ↑ | NA | |
| Glyphosate | *Conyza canadensis* | Resistant | = | ↓ | = | Feng et al. [29] |
| | | Susceptible | = | ↑ | = | |
| Glyphosate | *Conyza canadensis* | Resistant | = | ↓ | NA | Koger and Reddy [30] |
| | | Susceptible | = | ↑ | NA | |
| Glyphosate | *Lolium rigidum* | Resistant | = | = | = | Feng et al. [31] |
| | | Susceptible | = | = | = | |
| Glyphosate | *Lolium rigidum* | Resistant | = | ↓ | NA | Yeboah et al. [32] |
| | | Susceptible | = | ↑ | NA | |
| MSMA | *Xanthium strumarium* | Resistant | = | = | NA | Keese and Camper [33] |
| | | Susceptible | = | = | NA | |
| Chlorsulfuron | *Kochia scoparia* | Resistant | = | = | = | Saari et al. [34] |
| | | Susceptible | = | = | = | |
| 2,4-D | *Glechoma hederacea* | Tolerant | ↓ | = | NA | Kohler et al. [35] |
| | | Susceptible | ↑ | = | NA | |
| 2,4-D | *Raphanus raphanistrum* | Resistant | = | ↓ | = | Goggin et al. [6] |
| | | Susceptible | = | ↑ | = | |

| Herbicide | Weed | Biotype | Absorption | Translocation | Metabolism | Reference |
|---|---|---|---|---|---|---|
| Paraquat | *Crassocephalum crepidioides* | Resistant | = | ↓ | = | Ismail et al. [36] |
| | | Susceptible | = | ↑ | = | |
| Paraquat | *Hordeum leporinum* | Resistant | = | ↓ | NA | Preston et al. [7] |
| | | Susceptible | = | ↑ | NA | |
| Bispyribac sodium | *Echinochloa crus-galli* | Resistant | = | ↓ | NA | Riar et al. [8] |
| | | Susceptible | = | ↑ | NA | |
| Imazamox | *Echinochloa crus-galli* | Resistant | = | ↓ | NA | Riar et al. [8] |
| | | Susceptible | = | ↑ | NA | |
| Penoxsulam | *Echinochloa crus-galli* | Resistant | = | = | NA | Riar et al. [8] |
| | | Susceptible | = | = | NA | |
| Propoxycarbazone Sodium | *Bromus tectorum* | Resistant | = | = | ↑ | Park et al. [37] |
| | | Susceptible | = | = | ↓ | |

(=) % absorbed of the total applied radioactivity, % translocate of the total absorbed, and amount of metabolites formed were similar in resistant and susceptible weeds. (↓) values were lower in this biotype. (↑) values were higher in this biotype. NA: non-available.

**Table 1.** Absorption, translocation, and metabolism of $^{14}C$-herbicides in resistant and susceptible weeds.

http://dx.doi.org/10.5772/68092

are necessary to elucidate the mechanism of glyphosate resistance in weed population [28]. However, Feng et al. [29] suggested that glyphosate resistance is likely due to altered cellular distribution that impaired phloem loading and plastidic import of glyphosate resulting in reduced overall translocation as well as inhibition of EPSPS. Taken together, these results suggest that metabolic deactivation is not a likely mechanism for glyphosate resistance in weeds.

## 7. Radiation safety orientation

The purpose of radiation safety orientation is to protect researchers, employees, students, and the general public from overexposure to radiation. In that matter, it will be necessary to comply with regulations, laws, and guidelines regarding the safe use of radioactive material, such as $^{14}C$-herbicide.

It is mandatory that personal involved with the handling of radioactive material must attend to a training of radiological protection (RP), given by professionals certified by the regulatory agencies of each country.

The training should aim to achieve the clear and convincing transfer of the knowledge and recommendations on the subject. The main objective is to avoid deterministic health effects and to reduce the probability of stochastic health effects of ionizing radiation. For annual limits of exposure to ionizing radiation check the annals of the International Commission on Radiological Protection (ICRP) [26].

When handling a radiolabeled $^{14}C$-herbicide, the orientation for individual protection is to wear a Personal Protective Equipment (PPE), which consists of: laboratory coat exclusive for radiolabeled material handling, disposable plastic gloves, and protective goggles.

For general protection, the use of the international symbol of radioactive material is mandatory in every room or equipment where radiolabeled material is handled or stored, and only authorized personal should be allowed.

It is mandatory to have a radiation detector (usually Geiger-Müller) that must be turned on when handling radiolabeled material and the surface where it will be handled should be covered with an impermeable plastic film in order to prevent equipment contamination.

## 8. Radioactive waste management

The use of $^{14}C$-herbicide generates some waste that can be in the form of liquid scintillation vials, refuse, and biological waste. The volumes of the waste generated in research activities using $^{14}C$-herbicide are much smaller than those generated by reactor and fuel reprocessing operations; however, it still needs to be managed if the activity is superior of a certain threshold. This threshold will depend solely on which state of matter the waste is presented.

In Brazil, the *Comissão Nacional de Energia Nuclear* (CNEN) determines that if the solid waste activity is above 1 × $10^4$ kBq/kg (Norm CNEN-NN-8.01, 2014) [38], it must be stored on a flask specific for radioactive solid waste storage with the international radioactive symbol. If the activity is below the same value, it can be discarded as common waste.

The liquid waste generated by the utilization of $^{14}$C-herbicide is usually in the form of scintillation solution, and since the organic solvent used in the scintillation solution is not only toxic but also water insoluble, all the radiolabeled scintillation solution must be considered radioactive waste.

Every radioactive waste must be identified with all the information about the radionuclide, including: activity, volume, physical and chemical properties.

## 9. Conclusion

Absorption, translocation, and metabolism of herbicides are dependent upon active ingredient form and sensitivity of the target weed species. There is the need of further disclosure within the scientific community connected to the study of weeds regarding the use of $^{14}$C-herbicides on absorption, translocation, and metabolism studies in resistant and susceptible weed, mainly in the Brazilian conditions. In this chapter, a step-by-step methodology was suggested in order to meet this need, including the radiation safety orientation and management of resulting radioactive waste from the studies conducted in the laboratory. Techniques that use $^{14}$C such as tracers are extremely useful to study the herbicides behavior in the resistant weed, since the radiometric techniques offer the possibility of accurately determining very small amounts in a relatively short time. However, mechanism of resistance to herbicides in this resistant weed population compared with the susceptible population cannot be due to differential absorption, translocation, or metabolism of herbicide in weed; so other studies are necessary to elucidate the mechanism of herbicide resistance on weed population.

## Author details

Kassio Ferreira Mendes[1*], Ricardo Ferraz Silveira[1], Miriam Hiroko Inoue[2] and Valdemar Luiz Tornisielo[1]

*Address all correspondence to: kassio_mendes_06@hotmail.com

1 Ecotoxicology Lab, Center of Nuclear Energy in Agriculture (CENA), University of São Paulo (USP), Piracicaba, SP, Brazil

2 Departament of Agronomy, University of Mato Grosso State (UNEMAT), Tangará da Serra, MT, Brazil

## References

[1] Silva JF, Silva JF, Ferreira LR, Ferreira FA. Herbicidas: Absorção, translocação, metabolismo, formulação e misturas. In: Silva AA, Silva JF, editors. Tópicos em manejo de plantas daninhas. 1st ed. Viçosa, MG, Brazil: Universidade Federal de Viçosa; 2007. pp. 149-180.

[2] Heap I. The International Survey of Herbicide Resistant Weeds [Internet]. 2017. Available from: http://www.weedscience.org [Accessed: January 15, 2017]

[3] Beffa R, Figge A, Lorentz L, Hess M, Laber B, Ruiz-Santaella JP, Strek H. Weed resistance diagnostic technologies to detect herbicide resistance in cereal growing areas. A review. Julius-Kühn-Archiv. 2012;**434**:75-80. DOI: 10.5073/jka.2012.434.008

[4] Monquero PA, Christoffoleti PJ, Osuna MD, De Prado RA. Absorção, translocação e metabolismo do glyphosate por plantas tolerantes e suscetíveis a este herbicida. Planta Daninha. 2014;**22**(3): 445-451. DOI: 10.1590/S0100-83582004000300015

[5] Menendez J, Rojano-Delgado MA, Prado R. Differences in herbicide uptake, translocation, and distribution as sources of herbicide resistance in weeds. In: Myung K, Satchivi NM, Kingston CK, editors. Retention, Uptake, and Translocation of. 1st ed. Washington, DC, USA: ACS Symposium Series, American Chemical Society; 2014. pp. 141-157. DOI: 10.1021/bk-2014-1171.ch007

[6] Goggin DE, Cawthray GR, Powles SB. 2,4-D resistance in wild radish: Reduced herbicide translocation via inhibition of cellular transport. Journal of Experimental Botany.2016;**67**(11):3223-3235. DOI: 10.1093/jxb/erw120

[7] Preston C, Soar CJ, Hidayat I, Greenfield KM, Powles SB. Differential translocation of paraquat in paraquat-resistant populations of *Hordeum leporinum*. Weed Research. 2005;**45**(4): 289-295. DOI: 10.1111/j.1365-3180.2005.00454.x

[8] Riar DS, Norsworthy JK, Srivastava V, Nandula V, Bond JA, Scott RC. Physiological and molecular basis of acetolactate synthase-inhibiting herbicide resistance in barnyardgrass (*Echinochloa crusgalli*). Journal of Agricultural Food and Chemistry. 2013;**61**(2):278–289. DOI: 10.1021/jf304675j

[9] Powles SB, Yu Q. Evolution in action: plants resistant to herbicides. Annual Review of Plant Biology. 2010;**61**:317-347. DOI: 10.1146/annurev-arplant-042809-112119

[10] Shaner DL. Role of translocation as a mechanism of resistance to glyphosate. Weed Science. 2009;**57**:118-123. DOI:: 10.1614/WS-08-050.1

[11] Bromilow RH, Chamberlain K, Evans AA. Physicochemical aspects of phloem translocation of herbicides. Weed Science. 1990;**38**(3):305-314.

[12] Ferreira EA, Galon L, Aspiazú I, Silva AA, Concenço G, Silva AF, Oliveira JA, Vargas L. Glyphosate translocation in hairy fleabane (*Conyza bonariensis*) biotypes. Planta Daninha. 2008;**26**(3):637-643. DOI: 10.1590/S0100-83582008000300020

[13] Yu Q, Abdallah I, Han H, Owen M, Powles S. Distinct non-target site mechanisms endow resistance to glyphosate, ACCase and ALS-inhibiting herbicides in multiple herbicide resistant *Lolium rigidum*. Planta. 2009;**230**(4):713-723. DOI: 10.1007/s00425-009-0981-8

[14] Ghanizadeh H, Harrington KC, James TK. Genetic inheritance of restricted herbicide translocation in a glyphosate resistant *Lolium perenne* population. New Zealand Journal of Agricultural Research. 2016;**59**(3):269-279. DOI: 10.1080/00288233.2016.1189438

[15] Ge X, D'Avignon DA, Ackerman JJH, Sammons RD. In vivo $^{31}$P-nuclear magnetic resonance studies of glyphosate uptake, vacuolar sequestration, and tonoplast pump activity in glyphosate-resistant horseweed. Plant Physiology. 2014;**166**(3):1255-1268. DOI: 10.1104/pp.114.247197

[16] Ge X, D'Avignon DA, Ackerman JJH, Collavo A, Sattin M. Vacuolar glyphosate-sequestration correlates with glyphosate resistance in ryegrass (*Lolium* spp.) from Australia, South America, and Europe: a $^{31}$P NMR investigation. Journal of Agricultural and Food Chemistry. 2012;**60**(5):1243-1250. DOI: 10.1021/jf203472s

[17] IRSN – Institut de Radioprotection et de Sûreté Nucléaire. Carbon-14 and The Environment [Internet]. 2017. Available from: http://www.irsn.fr/EN/Research/publications-documentation/radionuclides-sheets/environment/Pages/carbon14-environment.aspx [Accessed: January 17, 2017].

[18] L'annunziata MF, editor. Handbook of Radioactivity Analysis. 3rd ed. San Diego: Academic Press; 2012. 1379 p.

[19] Coughtrey PJ, Nancarrow DJ, Jackson D. Extraction of carbon-14 from biological samples by wet oxidation. Communications in Soil Science and Plant Analysis. 1986;**17**(4): 393-399. DOI: 10.1080/00103628609367721

[20] Nandula VK, Vencill WK. Herbicide absorption and translocation in plants using radioisotopes. Weed Science. 2015;**63**(sp1):140-151. DOI: 10.1614/WS-D-13-00107.1

[21] Mendes KF, Martins BAB, Reis FC, Dias ACR, Tornisielo VL. Methodologies to study the behavior of herbicides on plants and the soil using radioisotopes. Planta Daninha. 2017;**35**:1-21.

[22] Kniss AR, Vassios JD, Nissen SJ, Ritz C. Nonlinear regression analysis of herbicide absorption studies. Weed Science. 2011;**59**(4):601-610. DOI: 10.1614/WS-D-11-00034.1

[23] Lovelace ML, Talbert RE, Hoagland RE, Scherder EF. Quinclorac absorption and translocation characteristics in quinclorac- and propanil-resistant and -susceptible barnyardgrass (*Echinochloa crus-galli*) biotypes. Weed Technology.2007;**21**(3):683-687. DOI: 10.1614/WT-06-060.1

[24] Wehtje G, Miller ME, Grey TL, Brawner Jr WR. Comparisons between X-ray film- and phosphorescence imaging-based autoradiography for the visualization of herbicide translocation. Weed Technology. 2007;**21**(4):1109-1114. DOI: 10.1614/WT-06-152.1

[25] Tranel P. Non-target-site Mechanisms of Herbicide Resistance-changes for Weed Management [Internet]. 2014. Available from: https://nomato.files.wordpress.com/2014/09/non-target-site-mechanisms-of-herbicide-resistance.pdf [Accessed: January 20, 2017].

[26] ICRP – International Commission on Radiological Protection. Annals of the ICRP – Occupational Intakes of Radionuclides Part 1 [Internet]. 2017. Available from: http://www.icrp.org/docs/Occupational_Intakes_P1_for_consultation.pdf [Accessed: January 16, 2017].

[27] Valenzuela JMV, Lownds NK, Sterling TM. Clopyralid uptake, translocation, metabolism, and ethylene induction in picloram-resistant yellow starthistle (*Centaurea solstitialis* L.). Pesticide Biochemistry and Physiology. 2001;**71**:11-19. DOI: 10.1006/pest.2001.2560

[28] Pérez S, Alister C, Kogan M. Absorption, translocation and allocation of glyphosate in resistant and susceptible Chilean biotypes of *Lolium multiflorum*. Weed Biology and Management. 2004;**4**(1):56-58. DOI: 10.1111/j.1445-6664.2003.00117.x

[29] Feng PCC, Tran M, Chiu T, Sammons RD, Heck GR, CaJacob CA. Investigations into glyphosate-resistant horseweed (*Conyza canadensis*): retention, uptake, translocation, and metabolism. Weed Science.2004;**52**(4):498-505. DOI: 10.1614/WS-03-137R

[30] Koger CH, Reddy KN. Role of absorption and translocation in the mechanism of glyphosate resistance in horseweed (*Conyza canadensis*). Weed Science. 2005;**53**(1):84-89. DOI: 10.1614/WS-04-102R

[31] Feng PCC, Pratley JE, Bohn JA. Resistance to glyphosate in *Lolium rigidum*. II. Uptake, translocation, and metabolism. Weed Science. 1999;**47**(4):412-415.

[32] Yeboah PA, Malone JM, Gill G, Preston C. Reduced glyphosate translocation in two glyphosate-resistant populations of rigid ryegrass (*Lolium rigidum*) from Fence Lines in South Australia. Weed Science. 2014;**62**(1):4-10. DOI: 10.1614/WS-D-13-00072.1

[33] Keese RJ, Camper ND. Uptake and translocation of [$^{14}C$] MSMA in cotton and MSMA-resistant and -susceptible cocklebur. Pesticide Biochemistry and Physiology. 1994;**49**(2):138-145. DOI: 10.1006/pest.1994.1041

[34] Saari LL, Cotterman JC, Primiani MM. Mechanism of sulfonylurea herbicide resistance in the broadleaf weed, *Kochia scoparia*. Plant Physiology. 1990;**93**(1):55-61. DOI: 10.1104/pp.93.1.55

[35] Kohler EA, Throssell CS, Reicher ZJ. 2,4-D rate response, absorption, and translocation of two ground ivy (*Glechoma hederacea*) populations. Weed Technology. 2004;**18**(4):917-923. DOI: 10.1614/WT-03-089R1

[36] Ismail BS, Chuah TS, Khatijah H. Metabolism, uptake and translocation of $^{14}C$-paraquat in resistant and susceptible biotypes of *Crassocephalum crepidioides* (Benth.) S. Moore. Weed Biology and Management. 2001;**1**(3):176-181. DOI: 10.1046/j.1445-6664.2001.00025.x/pdf

[37] Park KW, Fandrich L, Mallory-Smith CA. Absorption, translocation, and metabolism of propoxycarbazone-sodium in ALS-inhibitor resistant *Bromus tectorum* biotypes. Pesticide Biochemistry and Physiology. 2004;**79**(1):18-24. DOI: 10.1016/j.pestbp.2003.11.002

[38] CNEN - Comissão Nacional de Energia Nuclear. Gerência de rejeitos radioativos de baixo e médio níveis de radiação. Norma CNEN NN 8.01, Resolução CNEN 167/14, 30 de abril de 2014 [Internet]. 2014. Available from: http://appasp.cnen.gov.br/seguranca/normas/pdf/Nrm801.pdf. [Accessed: January 5, 2017].